A Handbook of Small-scale Energy Technologies

Praise for the book

'Practical Action is to be congratulated on bringing all its experience of decentralized energy systems into a single, easily accessible handbook just at a time when "energy access" is back on the aid agenda.'

Andrew Barnett, Director, The Policy Practice, Brighton

'This handbook provides a comprehensive review of small-scale decentralized options from renewable energy sources, benefiting poor people in developing countries. The chapters show how small-scale decentralized options are being deployed in rural areas of developing countries. The handbook is a useful tool for practitioners, students and other stakeholders involved in promoting and implementing small-scale energy technologies derived from sustainable energy resources.'

Dr Smail Khennas, senior energy and climate change expert, UK

'Decentralized energy systems are critical in bridging the energy access gap and in improving people's livelihoods. This handbook provides an excellent overview of various decentralized energy solutions and existing technological options that can bring energy to some of the world's most impoverished areas.'

Bahareh Seyedi, Energy Policy Specialist, UNDP

A Handbook of Small-scale Energy Technologies
Practical Answers

Edited by Neil Noble

Practical Action Publishing Ltd
The Schumacher Centre
Bourton on Dunsmore, Rugby,
Warwickshire CV23 9QZ, UK
www.practicalactionpublishing.org

Copyright © Practical Action, 2012

ISBN 978 1 85339 769 1 Hardback
ISBN 978 1 85339 770 7 Paperback
ISBN 978 1 78044 779 8 Library Ebook
ISBN 978 178044 770 4 Ebook

Neil Noble, ed. (2012) *A Handbook of Small-scale Energy Technologies: Practical Answers*, Practical Action Publishing, Rugby, UK.

All rights reserved. No part of this publication may be reprinted or reproduced or utilized in any form or by any electronic, mechanical, or other means, now known or hereafter invented, including photocopying and recording, or in any information storage or retrieval system, without the written permission of the publishers.

A catalogue record for this book is available from the British Library.

The contributors have asserted their rights under the Copyright Designs and Patents Act 1988 to be identified as authors of their respective contributions.

Since 1974, Practical Action Publishing (formerly Intermediate Technology Publications and ITDG Publishing) has published and disseminated books and information in support of international development work throughout the world. Practical Action Publishing is a trading name of Practical Action Publishing Ltd (Company Reg. No. 1159018), the wholly owned publishing company of Practical Action. Practical Action Publishing trades only in support of its parent charity objectives and any profits are covenanted back to Practical Action (Charity Reg. No. 247257, Group VAT Registration No. 880 9924 76).

Typeset by Bookcraft Ltd, Stroud, Gloucestershire

Printed in the UK

Contents

Photos vi
Figures vii
Tables and boxes viii

1	Introduction	1
2	Solar photovoltaic energy	6
3	Solar photovoltaic water pumping	18
4	Solar thermal energy	29
5	Solar drying	40
6	Solar distillation	48
7	Micro hydropower	58
8	River turbines	68
9	Biogas	73
10	Liquid biofuels and sustainable development	82
11	Biomass as a solid fuel	93
12	Charcoal production	104
13	Energy for rural communities	113
14	Refrigeration for developing countries	122
15	Rural lighting	133

Photos

La Encanada infocentre, Peru	12
Coconut drying in Bangladesh	34
A small solar tent dryer, Ghana	43
A Hohenheim dryer, Ghana	44
The Tunga–Kabiri micro-hydropower scheme, Kenya	59
A river current turbine in Peru	69
The fan belt and permanent magnet generator	71
Women design and manufacture improved cookstoves	101
Using a portable metal kiln to produce charcoal in Sudan	108
Low-wattage electrical cookers in Nepal	120
A biogas-powered lamp in Bandarawella, Sri Lanka	135
Testing a solar WLED lantern in Nepal	136

Figures

2.1	Components of a typical off-grid PV system	10
3.1	Village water supply with solar pumping	20
3.2	Solar irrigation system	21
3.3	Submerged multistage centrifugal motor pumpset	23
3.4	Submerged pump with surface-mounted motor	23
3.5	Floating motor pump	24
3.6	Suction pumpsets	25
4.1	The angle of the earth to the sun changes throughout the year	30
4.2	Dispersion of solar irradiance through the atmosphere	31
5.1	The Brace solar cabinet dryer	42
5.2	Section through the cabinet dryer	42
5.3	Tent dryer	43
5.4	Solar tunnel dryer layout	44
6.1	A single-basin still	49
6.2	Examples of solar still designs.	50
6.3	An inverted absorber solar still	53
7.1	Layout of a typical micro-hydro scheme	60
7.2	An impulse turbine	63
7.3	A reaction turbine	63
9.1	Fixed dome digester	77
9.2	Floating cover digester	77
9.3	Compact biogas digester	78
10.1	Global biofuel production figures	83
10.2	2010 ethanol production by region	83
10.3	2010 biodiesel production by region	84
10.4	Bioethanol production process	86
10.5	Biodiesel production process	87
11.1	A traditional metal stove and the improved Kenya Ceramic Jiko	96
11.2	Improved charcoal kiln found in Brazil, Sudan and Malawi	98
14.1	Passive/evaporative coolers	123
14.2	Sorption refrigerator	123
14.3	Compression refrigerator	124
14.4	Temperatures for safe storage	125
14.5	Sorption cooler	127
15.1	A simple solar system for lighting	138

Tables

3.1	Comparison of pumping techniques	19
3.2	Suitability of irrigation methods for use with solar pumps	20
5.1	The effect of air temperature on relative humidity	41
7.1	Classification of hydropower by size	59
7.2	Classification of turbine types	62
9.1	Biogas fuel equivalents	79
11.1	Efficiencies of some biomass energy conversion systems	95
14.1	Comparing refrigeration systems	130
15.1	Lighting capacity	137

Boxes

2.1	Sagar Island – solar island	7
2.2	Solar co-operative for Bangladeshi women	13
2.3	Solar energy to meet basic needs in the Himalayas	14
2.4	Glowstar – Kenya	14
11.1	Local manufacture of improved stoves	97
11.2	Dung as a traditional fuel	99
12.1	Charcoal from mesquite shrubs in Sudan	108
12.2	Biomass wastes for charcoal in Kenya	109

Chapter 1

Introduction

Energy is an essential part of modern life but many people in the world, especially the poorest people living in developing countries, do not have access to adequate energy resources. They may have no centralized electricity or the supply may be erratic and unreliable. This is why decentralized power delivery is needed, whether provided through modern technologies such as solar photovoltaic panels or through more traditional technologies which can be improved to deliver better performance, such as charcoal production.

In the domestic setting in developing countries energy is primarily used for lighting, cooking and heating, and to power small appliances such as radios. Families that can acquire more efficient sources of power will have a better living environment. Solar lighting can replace kerosene lamps, making the home lighter for longer as well as safer and cleaner. Despite the initial costs, replacing older forms of power generation can ultimately save money for families who would otherwise be spending money on fuel on a weekly basis. Access to energy enables people to undertake income-generating activities that would otherwise not be possible. Even small amounts of energy can make a big difference in people's everyday lives and in the amount of money they can earn.

In rural areas energy is needed for agriculture and crop processing. Energy for water pumping helps to increase crop production through improved irrigation. Small-scale milling or grinding equipment can greatly improve the income of farmers and allow farming families to have some control over how they manage their crops. Powered oil-expelling equipment will increase the processing capacity.

Energy is not only required for agriculture; urban areas need energy and often do not have sufficient supplies. Small businesses and shops need energy. Hairdressers in Kenya make the most of the energy delivered through renewable sources to improve their service to customers and help increase their income.

Energy is also needed to power the communication revolution, allowing people to have access to mobile phones which bring a range of benefits, keeping families in touch with one another and improving income generation through better market awareness by allowing people to know the best time to sell their produce and where the best resources can be found.

In schools, electric lighting and computers help children learn, with access to the internet providing them the opportunity to know more about the wider world. Hospitals and health clinics cannot function effectively without energy. Importantly vaccines must be refrigerated to ensure they remain effective.

Although centralized energy delivery is growing in almost every part of the world, there remains a large unmet demand for energy and delivery is very uneven. Grid connection will not reach sufficient numbers of people in forthcoming years and supply from centralized grid systems will remain unreliable as demand outstrips supply. Decentralized energy delivery has the advantage that it can spread rapidly to remote areas and is economically more attainable. This means that the technologies described in this publication will be suitable for the many people who will otherwise be unable to access an energy supply.

This book brings together a selection of the technical briefs produced by Practical Action on the subject of energy. The work of Practical Action and others has been to enable people living in poverty to improve their living standards. One way of doing this is by improving access to energy. Small-scale renewable energy technologies are a practical way of delivering clean energy to rural populations, providing energy for lighting, radios and other small domestic appliances as well as supporting economic activities that will improve incomes. Access to energy can also be a driver for poverty reduction by enabling businesses to develop and production to grow. This may be through a village-scale initiative, such as a micro-hydro scheme which could supply energy to many people and their businesses through a mini grid, but it also includes stand-alone systems that are used by one household independently.

Not only do clean energy technologies improve the lives of people, they also help reduce environmental pollution by replacing fossil fuels. Consequently, some small-scale renewable energy projects have been financed by carbon credit schemes.

Solar energy

Solar energy can be captured in two ways. The more traditional approach is to capture heat using some sort of thermal device such as solar water heater, while the other is to generate electricity directly through the use of photovoltaic panels.

Solar photovoltaic energy is ideal for household systems or isolated applications such as solar water pumping. These systems are relatively easy to transport and install, and require very little operation and maintenance. Although running costs can be very low, for many people solar photovoltaic technology is still an expensive option in terms of the initial capital cost. However, solar photovoltaic technologies have been falling in price over the past few years and are likely to continue to fall even further in coming years. The price of solar panels fell by almost 50 per cent in 2011 according to Bloomberg New Energy Finance (BNEF). The overall cost of solar electricity is now cheaper than that produced by diesel generators when looking at periods of seven years or longer.

The falling cost of solar photovoltaic technology components is mainly a result of the economies of scale arising from increased production, rather than any radical changes in the technology. Commercial panels still only convert 15 to 18 per cent of the energy that reaches them, but this low efficiency is relatively unimportant compared to the cost. This reduction in cost means that the technology is becoming available to poorer households.

Small-scale solar thermal energy options are also discussed in this publication, concentrating specifically on solar drying and solar distillation. Solar thermal energy has some significant advantages over the more technologically advanced PV systems in that the initial investment can be very small and the running costs are also minimal. The technology can be manufactured locally on a small scale. However, many small-scale solar thermal energy technologies have not dramatically changed over the years and have, to some extent, been relatively neglected in comparison to the developments in solar PV. Solar water heating is perhaps the most widely used solar thermal technology. It has become common in some regions such as the Mediterranean and North Africa, while not much used in other regions. To a lesser extent solar water heating has progressed in a similar fashion to solar PV in that it has developed and expanded in its use with steadily increasing production and thus reducing in cost. The technology is suitable for small-scale local manufacture.

Other solar thermal technologies include solar food drying, which Practical Action has promoted over many years. Fruit and herbs can be dried and sold in the off season, helping to bolster the farmer's income.

Micro hydro and river turbines

Micro-hydro schemes are usually installed at the village level, so that one scheme provides electricity to many homes and to industries within the village. The investment needed for a micro-hydro scheme is large compared to household income, so financial structures play an important role in the implementation of micro-hydro schemes. Keeping the cost down to a realistic level is an important factor in making this technology viable. Cost can vary considerably depending on the physical locations and the terrain of the site. Smaller sites tend to be more expensive per kilowatt of power delivered and cost tends to increase as the head (i.e. the height that the water will fall through) decreases.

Micro hydro has a significant history in some regions of the world, while in other parts its potential has not been realized. The technology is well proven and reliable. Experience is being gained in new regions of the world, especially in Africa where small-scale community hydro is beginning to establish a firm footing.

River turbines are different from conventional hydro schemes in that there is no head of water. Instead the energy in the flow of the water is used to power a turbine in a similar way to air flow in wind turbines. This technology is not widely used but they have been of increasing interest not only for developing countries but also as part of the energy solution for industrialized countries. Practical Action has developed this approach in Peru while larger systems have been developed in North America and Europe. The technology is closely related to those used to harness tidal energy.

Biomass, biogas, and biofuels

Biomass as a solid fuel includes a range of materials. Most familiar is wood used for cooking in many households, while others use crop residues such as rice husk either as fine residue or made into briquettes. Processing is very important and small changes to traditional approaches can provide significant improvements in quality and quantity.

Biomass is used in one form or another by the majority of people in the world, primarily as a fuel for cooking either as wood or as charcoal made from wood. However, there are many options in the use of biomass and significant research has been undertaken in some areas of biomass processing, especially in the production of liquid biofuels that can be used to replace fossil fuels. Although large-scale biofuel production has generated controversy because it competes for land that would be used to produce food, small-scale biofuel production can offer local producers some significant benefits.

Charcoal production is done in many parts of the world on an informal basis and charcoal is widely used in urban areas as a cooking fuel. Improvements to the basic approach to production can significantly reduce emissions and improve the output of the kilns. In recent years there has been interest in developing charcoal production using crop waste and invasive plant species as a fuel source instead of wood. This is highlighted in Sudan where mesquite has become a real problem, but can be used to produce charcoal which is sold in the cities.

Liquid biofuels include a range of substances such as biodiesel and bioethanol. The expansion of liquid biofuels has been rapid in recent years and will likely continue to rise as attempts are made to replace fossil fuels. Large-scale operations are controversial because they occupy land that has the potential to produce food crops. Small-scale biofuel operations have less controversy associated with them and can make a positive impact for communities in developing countries

Biogas has been promoted over a number of years and can work well supplying energy to homes for cooking and lighting. It can also be used to generate electricity if the supply is sufficiently large. It has the advantage of providing a means of managing waste materials in sanitation sites. Bio-sanitation is of great interest as a sanitation solution in many locations, especially in urban areas where more simple toilets requiring more space or regular emptying can be substituted.

Energy for rural communities

The issues surrounding energy in rural communities are not just about the fuels that are available but also how to make best use of them and what they can be used on. Small-scale energy supplies to domestic homes will be used for lighting and cooking; in some colder regions heating will also be important. Small appliances such as radios are popular along with mobile phone charging.

Away from the domestic setting, energy is needed for schools and health centres where computers or refrigeration for vaccines require a reliable energy supply. Energy can also be used to generate income. Micro-hydro schemes have prompted the development of small-scale industry as a way of making use of the energy available in daylight hours, when domestic use is at its lowest.

Delivering electricity to more remote households is a challenge that is often regarded as too difficult or costly to achieve, but there are low-cost options available which are looked at within this book. However, it is not only electrical energy that needs to be considered. Often there will be competing energy options that can potentially be used so it is important to be aware of how these can be used and what the relative merits are. Two areas where this can be seen are food preservation, where options include coolers and refrigerators, and lighting, where options range from candles to modern light-emitting diode systems.

About Practical Answers

Practical Action has worked on providing information on appropriate technologies throughout its existence and one aspect of this is Practical Answers, its technical information service. The service aims to provide information and understanding about a wide range of low-cost appropriate technology options that can be used to help people move out of poverty. It covers such diverse topics as agriculture, low-cost construction, disaster risk reduction, food processing, water and sanitation, small-scale manufacturing and production, transport, waste management, as well as energy technologies covered in this publication and others that have not been included such as wind technology and improved stove design – although these subjects are discussed in Chapters 8 and 15 (wind), and Chapter 11 (stove design).

Technical briefs, manuals and other formats such as videos and audio files have been produced to help guide people to make technology choices and find the best route to better lives, either directly or through intermediaries such as development charities like Practical Action. This information is available through the Practical Action website, on CD, and in print. Individuals can also make direct enquiries to any of the Practical Action offices on any of these technical subjects. Questions can be submitted free of charge to any office where Practical Action can call upon its staff and partners to provide an answer that will be as relevant as possible to the particular circumstances.

This book has been produced as a collaboration between Practical Answers and Practical Action Publishing, to make Practical Action's technical information on energy available in printed form. It covers just some of the energy technologies that Practical Action (or, under its previous name, ITDG) has been working on over the years, and providing information on through its technical information service. The subjects included in this book have consistently been of interest, although the technology has developed over the years. These changes have been reflected in the document revisions.

Some well-established small-scale technologies such as hydro and solar have been included here, along with examples of how energy can be effectively used for such things as lighting and refrigeration. Topic areas such as wind power and stoves that have not been described in detail in this book are covered in other publications, and each chapter includes a bibliography where further reading is listed. Readers who are interested in finding out about other technologies can look at the Practical Action website (http://practicalaction.org/practicalanswers/).

Chapter 2
Solar photovoltaic energy

This chapter explains solar photovoltaic technology and provides some context to its use in developing countries, looking at the main components and at issues of cost. The different types of panels that are commercially available are outlined and how these are incorporated into a typical solar system. Typical applications for developing countries are presented, and the potential for local assembly and dissemination.

Keywords: solar, renewable, photovoltaic, hybrid systems, photovoltaic applications

Introduction

Photovoltaics (PV) is a technology that converts sunlight directly into electricity. It was discovered in 1839 by the French scientist Becquerel, who detected that when light was directed onto one side of a simple battery cell, the current generated could be increased. In the late 1950s, the space programme provided the impetus for the development of crystalline silicon solar cells; the first commercial production of PV modules for terrestrial applications began in 1953 with the introduction of automated PV production plants.

Today, PV systems have huge value in areas remote from an electricity grid where they can provide power for water pumping, lighting, vaccine refrigeration, electrified livestock fencing, telecommunications and many other applications. With the global demand to reduce carbon dioxide emissions, PV technology is also gaining popularity as a mainstream form of electricity generation. Millions of solar PV systems are currently in use worldwide, with an installed capacity of 40 GW globally by the end of 2010 (REN21, 2011), yet this is a tiny proportion of the vast potential that exists for PV as an energy source.

Photovoltaic modules provide an independent, reliable electrical power source at the point of use, making PV particularly suited to remote locations. However, solar PV is increasingly being used in homes and offices for electricity to replace or supplement grid power, often in the form of solar PV roof tiles. The daylight needed is free, but the cost of equipment can take many years to achieve payback. However, in remote areas where grid connection is expensive, PV can be the most cost-effective power source.

The use of PV electricity in developing countries

Most of the world's developing countries are within the tropics and hence have ample solar insolation (the total energy per unit area received from the sun).

The tropical regions also benefit from having only a small seasonal variation of solar insolation, even during the rainy season, which means that, unlike northern industrial countries, they can harness solar energy economically throughout the year. China, India and other developing countries are emerging as major solar PV manufacturers.

The dominant application for PV in developing countries is the solar home system (SHS). This involves the installation of PV systems of 30–50 peak watts (Wp), costing about US$300–500 (or £192–320, at an exchange rate of $1 = £0.64) each, in individual homes, mainly in rural areas. Apart from SHS, other applications of PV in developing countries include: (1) PV-powered remote telecommunications equipment; (2) rural health clinic refrigerators; (3) rural water pumping; (4) solar lanterns; and (5) PV battery-charging programmes, which allow rural residents to purchase or rent batteries to provide electricity to their homes and then recharge them at PV-powered charging stations. A few attempts have been made to establish PV-powered village power grids in developing countries, such as in Sagar 'Solar Island' off the coast of India (see Box 2.1).

Box 2.1 Sagar Island – solar island

Sagar Island is in the south-western corner of the Ganges delta, in India. The West Bengal Renewable Energy Development Agency (WBREDA) has been working on Sagar Island since 1996 to address the problem of energy supply. Since then it has set up a total of 11 small solar PV power plants on Sagar Island and its neighbour Maushuni Island. Each plant has its own mini-grid system that distributes electric power to the surrounding villages. The grids are switched on for six hours a day, from 6pm to midnight, and are managed by co-operative societies formed by the villagers who use the power.

The 11 power plants in operation supply stable and reliable 400 / 230 V, 3-phase, 50 Hz power for six to seven hours a day through local distribution lines. The combined capacity of the plants is 400 kW and WBREDA estimates that a further 400 kW is needed in order to electrify all the villages on the two islands.

Source: Ashden, 2003

Cost of solar PV

The process of producing efficient solar cells is costly due to the use of expensive pure silicon and the energy consumed, and cost has been the major barrier to the widespread uptake of PV technology. As materials technology improves, costs are slowly dropping, making PV technology more attractive.

There are also economies of scale so that the cost comes down as production increases. Every time the installed capacity has doubled the cost has dropped by 20 per cent per year: module prices have fallen from $30/Wp in 1975 to less than $1/Wp in 2012 (i.e. from around £19.20 to less than £0.64). Costs rose slightly in 2004 due to high demand (which outpaced supply) and the rising cost of silicon. The expectation is that the cost of PV will continue to come down as mass production increases and technologies evolve. (Cost of PV modules is usually given in terms of peak wattage (Wp), which is the power rating of the panel at peak conditions – that is, at 1 kWm^{-2} irradiance at 25°C.)

Technical aspects

The nature and availability of solar radiation is described in Chapter 4 'Solar thermal energy' (Noble, 2012). Once the solar energy reaches the surface of a photovoltaic cell, the electrons become energized in proportion to the intensity and spectral distribution (wavelength distribution) of the light. When their energy level exceeds a certain point a potential difference is established across the cell. This is then capable of driving a current through an external load, such as a light or radio.

PV modules

All modern, commercial PV devices use silicon as the base material, mainly as monocrystalline or multicrystalline cells, but more recently also in amorphous form. Other materials such as copper indium diselenide and cadmium telluride are being developed with the aim of reducing costs and improving efficiencies. A monocrystalline silicon cell is made from a thin wafer of a high purity silicon crystal, doped with a minute quantity of boron. Phosphorus is diffused into the active surface of the wafer. The front electrical contact is made by a metallic grid; the back contact usually covers the whole surface. An anti-reflective coating is applied to the front surface. Typical cell size is about 15 cm diameter.

The modules in a PV array are usually first connected in series to obtain the desired voltage; the individual strings are then connected in parallel to allow the system to produce more current. The modules are then protected by encapsulation between glass and a tough metal, plastic, or fibreglass back. This is held together by a stainless steel or aluminium frame to form a module. These modules, usually comprised of about 30 PV cells, form the basic building block of a solar array. Modules may be connected in series or parallel to increase the voltage and current, and thus achieve the required solar array characteristics that will match the load. Typical module size is 50 Wp and produces direct current electricity at 12 V (for battery charging, for example).

Solar cell types

Commercially available modules fall into three types based on the solar cells used.

Monocrystalline cell modules. The highest cell efficiencies of around 15–18 per cent are obtained with these modules. The cells are cut from a monocrystalline silicon crystal.

Multicrystalline cell modules. The cell manufacturing process is lower in cost but cell efficiencies of only around 15 per cent are achieved. A multicrystalline cell is cut from a cast ingot of multicrystalline silicon and is generally square in shape.

Amorphous silicon modules. These are made from thin films of amorphous silicon where efficiency is much lower (10–12 per cent) but the process uses less material. The potential for cost reduction is greatest for this type and much work has been carried out in recent years to develop amorphous silicon technology. Unlike monocrystalline and multicrystalline cells, with amorphous silicon there is some degradation of power over time.

Solar arrays

An array can vary from one or two modules, with an output of 10 W or less, to a vast bank of several kilowatts or even megawatts.

Flat plate arrays, fixed at a tilted angle and facing towards the equator, are most common. The angle of tilt should be approximately equal to the angle of latitude for the site. A steeper angle increases the output in winter; a shallower angle means more output in summer. It should be at least 10 degrees, to allow for rain runoff.

Tracking arrays follow the path of the sun during the day and thus theoretically capture more sun. However, the increased complexity and cost of the equipment rarely make it worthwhile.

Mobile (portable) arrays can be of use if the equipment is required in different locations such as with some lighting systems or small irrigation pumping systems.

Solar PV systems

While in industrialized countries there has been a rapid increase in grid-connected PV systems, in developing countries the majority of PV systems are stand-alone off-grid systems. The off-grid systems can be used to drive a load directly; water pumping is a good example. Water is pumped during the hours of sunlight and stored for later use; or a battery can be used to store power for use for lighting during the evening. If a battery-charging system is used, electronic control apparatus will be needed to monitor the system. All the components other than the PV module are referred to as the balance-of-system (BOS) components. Figure 2.1 shows a typical configuration for an off-grid PV system. Such systems can often be bought as kits and installed by semi-skilled labour.

For correct sizing of PV systems, the user needs to estimate the demand on the system, as well as acquiring information about the solar insolation in the area (approximations can be made if no data is readily available). It is normally assumed that for each Wp of rated power the module should provide 0.85 watt hours of energy for each $kWhm^{-2}$ per day of insolation (Hulscher, 1994). Therefore if we consider a module rated at 200 Wp and the insolation for our site is 5 $kWhm^{-2}$ per day (typical value for tropical regions), then our system will produce 850 Wh per day (that is $200 \times 0.85 \times 5 = 850$).

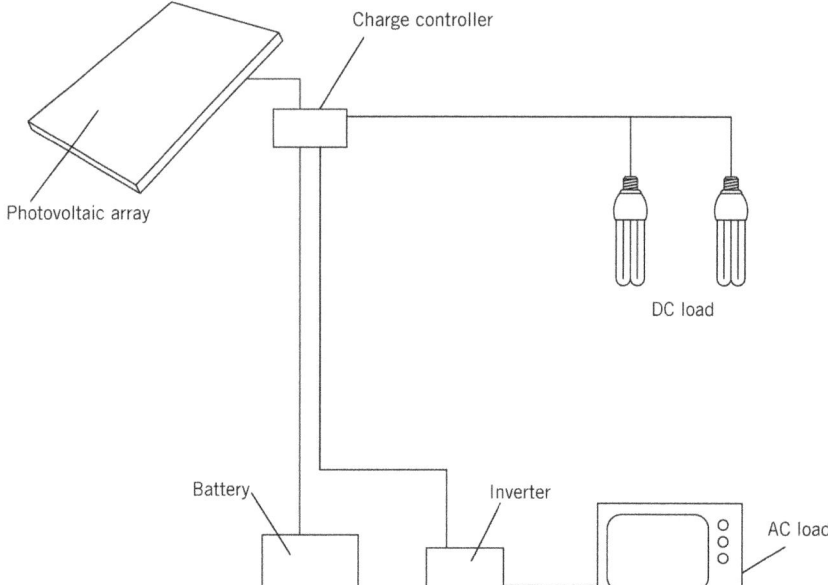

Figure 2.1 Components of a typical off-grid PV system

Source: Neil Noble/Practical Action

Some benefits of photovoltaics

- *No fuel requirements.* In remote areas diesel or kerosene fuel supplies are erratic and often very expensive. The recurrent costs of operating and maintaining PV systems are small.
- *Modular design.* A solar array comprises individual PV modules, which can be connected to meet a particular demand.
- *Reliability of PV modules.* This has been shown to be significantly higher than that of diesel generators.
- *Easy to maintain.* Operation and routine maintenance requirements are simple.
- *Long life.* With no moving parts and all delicate surfaces protected, modules can be expected to provide power for 15 years or more.
- *National economic benefits.* Reliance on fossil fuels such as coal and oil is reduced.
- *Environmentally benign* There is no pollution through the use of a PV system – nor is there any heat or noise generated which could cause local discomfort. PV systems bring great improvements in the domestic environment when they replace other forms of lighting – kerosene lamps, for example.

PV applications in less-developed countries

Rural electrification

- lighting and power supplies for remote buildings (mosques, churches, temples, etc; farms, schools, mountain refuge huts) – low-wattage fluorescent or LED lighting is recommended;
- power supplies for remote villages;
- street lighting;
- individual house systems (solar home systems);
- battery charging;
- mini grids.

See Chapter 15, 'Rural lighting' (Noble, 2012).

Water pumping and treatment systems

- pumping for drinking water;
- pumping for irrigation;
- dewatering and drainage;
- ice production;
- saltwater desalination systems;
- water purification.

See Chapter 3, 'Solar photovoltaic water pumping' (Noble, 2012).

Health care systems

- lighting in rural clinics;
- UHF transceivers between health centres;
- vaccine refrigeration;
- ice pack freezing for vaccine carriers;
- sterilizers;
- blood storage refrigerators.

PV is frequently used to power vaccine refrigeration in remote health centres. See the Practical Action Technical Brief, Solar Photovoltaic Refrigeration of Vaccines.

Communications

- radio repeaters;
- remote TV and radio receivers;
- remote weather measuring;
- mobile radios;
- rural telephone kiosks;
- data acquisition and transmission (for example, river levels and seismographs).

La Encanada infocentre, Peru has solar panels to generate electricity and a satellite for connectivity

Source: Practical Action / Jaime Soto

Transport aids

- road sign lighting;
- railway crossings and signals;
- hazard and warning lights;
- navigation buoys;
- road markers.

Security systems

- security lighting;
- remote alarm system;
- electric fences.

Miscellaneous applications

- ventilation systems;
- pumping and automated feeding systems on fish farms;

- solar water heater circulation pumps;
- boat and ship power;
- vehicle battery trickle chargers;
- earthquake monitoring systems;
- emergency power for disaster relief.

Local assembly of PV modules and BOS components

While generally it is only the larger developing countries which have capacity to manufacture solar cells, it is increasingly common for assembly of the module and the balance-of-system (BOS) components to be done in many developing countries. This not only reduces the overall cost of the system, but creates local employment and ensures that the systems are designed for local applications.

> **Box 2.2 Solar co-operative for Bangladeshi women**
>
> The consultancy Prokaushali Sangsad Limited (PSL) realized the need for both good quality lighting and employment for unskilled women on the remote island of Char Montaz in Bangladesh. They set up the Coastal Electrification and Women's Development Co-operative (CEWDC). The 35 co-operative members assemble photovoltaic solar home systems and sell them to island families. They also run a battery-charging service.
>
> Solar home systems are small, stand-alone electrical systems. They consist of a photovoltaic (PV) module; a rechargeable battery; a charge controller, which prevents the battery from being over-charged or deep-discharged; fluorescent lamps rated from 6 to 11 W; wiring and fixtures. The PV modules are rated at 20 to 80 Wp with 50 Wp the most popular size. A system based on a 20 Wp module can supply two or three 6 W lamps for about four hours per day: at the other end of the range, an 80 Wp system can power four 8 W lamps and a black and white television set.
>
> *Source:* Ashden, 2005

Dissemination to remote areas

When disseminating solar PV to remote locations, it is important that there is sufficient engineering capacity in the area to supply, install and maintain the solar systems. In addition, in low-income areas, microcredit is an important element of a dissemination programme, to allow the cost of the system to be paid back by sales over a period of time.

Hybrid systems

Solar PV can be used in conjunction with other energy technologies to provide an integrated, flexible system for remote power generation. These systems are referred to as hybrid systems. Common configurations of hybrid systems include a solar PV array, wind generator and diesel generator set which would allow generation in all weather conditions. Such systems need careful planning.

> **Box 2.3 Solar energy to meet basic needs in the Himalayas**
>
> Over the past 10 years a pioneering project has been introducing solar technology to remote and inaccessible villages in the Himalayas. Run by the Barefoot College in Rajasthan, India, the project has shown that, with appropriate training, poor and rural communities can install solar equipment in their villages and then maintain it without any further external help. The project has trained illiterate and semi-literate villagers as 'Barefoot Solar Engineers' (BSEs) at its Barefoot College. After the training, they return to their home villages to install solar units and provide their communities with a skilled and competent repair and maintenance service.
>
> *Source:* Ashden, 2003

Solar lanterns

Originally designed for the outdoor leisure market in Western countries, this simple lantern with a small PV module (5–10 watts) is ideal for use in rural areas of developing countries to replace kerosene lamps.

> **Box 2.4 Glowstar – Kenya**
>
> The solar lantern, 'Glowstar', has been designed as a low-cost alternative to a solar home system and is intended to allow rural families in Kenya to climb the first step on the 'energy ladder'. The lantern is cheap to maintain and harnesses a free and plentiful source of energy as it is powered by sunshine. The solar lantern kit consists of a photovoltaic panel and a lantern containing a high-efficiency lamp, a rechargeable battery and a charge control circuit. The solar lantern is ideal for any application where there is no local connection to grid electricity such as rural households and farms, schools and colleges, hospitals, health clinics and other community centres. It also has important applications where there is an inconsistent or unreliable supply of electricity.

LED lighting

In recent years solar PV has been coupled with light-emitting diodes (LEDs) to provide energy-efficient light. Recent advancements in LED technology have led to the development of white-light-emitting diodes (WLEDs). WLEDs provide a bright white light that is ideal for domestic lighting. The advantage of using LEDs with solar PV systems is that the LED requires a much lower wattage (less than conventional compact fluorescent light bulbs), therefore the size and the cost of the solar system are much reduced for each household.

Micro grids

Solar PV can be used for providing power for small grid systems, with centralized power generation. As the cost of PV cell production drops, their use for medium-scale electricity production is being adopted more widely. There is also scope for large-scale electricity production for such applications as peak power provision.

Further information

Technical Briefs

Batteries, http://practicalaction.org/batteries
Solar Photovoltaic System Design Info Sheet,
    http://practicalaction.org/solar-photovoltaic-system-design-info-sheet
Solar PV Refrigeration of Vaccines,
    http://practicalaction.org/solar-photovoltaic-refrigeration-of-vaccines-1

Bibliography

Ashden (2003) 'WBREDA, India: Solar powers island mini-grids', Trust Awards for Sustainable Energy, www.ashden.org/winners/wbreda [accessed 7 August 2012].

Ashden (2005) 'Solar cooperative for rural women', Awards for Sustainable Development, www.ashden.org/winners/psl.

Ashden (2008) 'Solar power in the Himalayas, Ladakh, Northern India' [blog] www.ashden.org/blog/solar-power-himalayas-ladakh-northern-india (posted 3 October) [accessed 7 August 2012].

Ashden (2009) 'Case study summary: Prokaushali Sangsad Ltd, Bangladesh', www.ashden.org/files/PSL%20Bangladesh%20case%20study%20full.pdf

Cabraal, A., Cosgrave-Davies, M. and Schaeffer, L. (1996) *Best Practices for Photovoltaic Household Electrification Programs,* World Bank, Washington, DC.

Foley, G. (1995) *Photovoltaic Applications in Rural Areas of the Developing World,* World Bank, Washington, DC.

Garg, H.P., Gouri, D. and Gupta, R. (1997) *Renewable Energy Technologies,* Indian Institute of Technology, Delhi, and the British High Commission, London.

Hankins, M. (2000) 'A case study on private provision of photovoltaic systems in Kenya', in *Energy and Development Report: Energy Services for the World's Poor,* ESMAP, chapter 11, World Bank, Washington, DC. Available at: www.worldbank.org/html/fpd/esmap/energy_report2000/

Hulscher, W. and Fraenkel, P. (1994) *The Power Guide: An International Catalogue of Small-scale Energy Equipment,* 2nd edn, Practical Action Publishing, Rugby, and TDG University of Twente, the Netherlands.

Karekezi, S. and Ranja, T. (1997) *Renewable Energy Technologies in Africa.* AFREPREN/SEI/Zed Books, London.

Louineau, Jean-Paul (2008) *A Practical Guide to Solar Photovoltaic Systems for Technicians: Sizing, Installation and Maintenance,* Practical Action Publishing, Rugby.

Louineau, J.P., Dicko, M., Fraenkel, P., Barlow R. and Bokalders, V. (1994) *Rural Lighting: A Guide for Development Workers*. Practical Action Publishing, Rugby, and Stockholm Environment Institute.

Markvart, Tomas (ed.) (2000) *Solar Electricity*, 2nd edn, John Wiley & Son, Hoboken, NJ.

Markvart, T. and Castañer, L. (eds) (2003) *Practical Handbook of Photovoltaics: Fundamentals and Applications*, Elsevier, Oxford.

Noble, N., (ed.) (2012) *A Handbook of Small-scale Energy Technologies: Practical Answers*, Practical Action Publishing, Rugby.

REN21 (2011) *Renewables 2011: Global Status Report*, Renewable Energy Policy Network for the 21st Century, Paris.

Useful addresses

International Solar Energy Society: www.ises.org
International Energy Agency Photovoltaic Power Systems Programme: www.iea-pvps.org/
Directory of renewable energy suppliers: http://energy.sourceguides.com/
Home Power magazine: www.homepower.com/
US National Centre for Photovoltaics: www.nrel.gov/ncpv
Solar Global (solar company directory): www.solar-global.net/solar-directory.html
Solarbuzz Inc., a directory of solar system suppliers: www.solarbuzz.com

Manufacturers and suppliers of photovoltaic products

Here are just a few examples of the many companies providing solar power around the world. This list of suppliers is not exhaustive and does not imply endorsement by Practical Action.

Alternative Technologies Pvt. Ltd
3 Canald Road, Graniteside, Harare, Mash Cent, Zimbabwe
Tel: +263 4 781 972-7

CIME Commercial S.A.,
132 Industrial Ave Lima – Tie, Peru
Tel: +51 1 326 0601
www.cime.com.pe/index.php

Link Intertrade (Private) Ltd
385C Old Kotte Road, Rajagiriya, Sri Lanka
Tel: +94 1 873 211-2

Lotus Energy Pvt. Ltd,
Bhatbhateni Dhunge Dhara, PO Box 9219, Kathmandu, Nepal
Tel: +977 1 418 203
www.lotusenergy.com

Kenital Solar Energy
Ngong Road, PO Box 19764, Nairobi, Kenya
Tel: +254 2 715 960
www.kenital.com

M/S Wisdom Solar (Private) Ltd,
No. 434, Thalawatugoda Road Madiwela, Kotte, Sri Lanka
Tel: +94 011 277 9790 1
www.wisdomsolar.lk/about-us.html

Solamatics
31 Edison Crescent, Graniteside, PO Box 2851, Harare, Zimbabwe
Tel: +263 4 749 930

Solar Power and Light Co. Ltd
10 Havelock Place, Colombo 5, Sri Lanka
Tel: +94 1 688 730

Solarman Co.
PO Box 11545, Khartoum, Sudan
Tel: +249 11 472 337

U.T.E. Group of Companies
PO Box 2074, Khartoum, Sudan
Tel: +249 11 70147

Solar modules

Bharat Heavy Electricals Ltd (BHEL)
BHEL House, Siri Fort, New Delhi 110049, India
Tel: +91 11 26001010 (multiple lines)
www.bhel.com

Central Electronics Ltd (CEL)
4 Industrial Area, Sahibabad, Uttar Pradesh 201010, India
Tel: +91 120 2895165
www.celsolar.com

Sharp Photovoltaics Div
282-1 Hajikami, Shinjo-cho, Kita-Katsuragi-gun, Nara Prefecture 639-2198, Japan
Tel: +81 745 63 35 63
www.sharp-solar.com/en/

Zentrale Bosch Solar Energy AG
Robert-Bosch-Strasse 1, 99310 Arnstadt, Germany
Tel: +49 (0)361 2195 0
www.bosch-solarenergy.de/

'Solar photovoltaic energy' was rewritten by Alison Doig for Practical Action in October 2007 and last updated in April 2012. Alison has a PhD in rural energy systems and spent eight years as an energy specialist for ITDG / Practical Action working on micro hydropower, cookstoves and energy policy. Since then she has worked for WWF and as a private consultant, and her current role is as policy adviser at Christian Aid, specializing in climate change, sustainable development and low-carbon energy.

Chapter 3
Solar photovoltaic water pumping

This chapter describes the particular application of solar technology and how it compares with other pumping methods. Some typical set-ups in developing countries are described, including the various components and the different pumps that can be used. Critical factors to a successful pumping system include assessing the water available and sizing the solar pump.

Keywords: solar, photovoltaic, renewable, energy, water pumping, photovoltaic applications

Introduction

Water pumping has a long history, and many pumping methods have been developed. People have used a variety of power sources, namely human energy, animal power, hydropower, wind and solar power, and fuels such as diesel for small generators. The most common pumps used in remote communities are:

- handpumps;
- direct-drive diesel-driven borehole pumps;
- electric submersible pumps with diesel generator;
- solar submersible pumps.

The relative merits of the various pumping methods are laid out in Table 3.1.

Applications

Solar pumps are used principally for three applications:

- village water supply;
- livestock watering;
- irrigation.

A solar pump for village water supply is shown schematically in Figure 3.1. The village will have a constant water demand, and it is necessary to store water for periods of low insolation (low solar radiation / days when it is not sunny). In environments where rainy seasons occur, some of this demand can be met by rainwater harvesting during the rainy season. Ideally in Sahelian Africa the storage would be sufficient for three to five days of water demand. In practice some installed tanks do not have sufficient capacity and are smaller than a day's

Table 3.1 Comparison of pumping techniques

Technique	Advantages	Disadvantages
Handpumps	Local manufacture is possible Easy to maintain Low capital cost No fuel costs	Loss of human productivity elsewhere Often an inefficient use of boreholes Low flow rates
Animal-driven pumps	More powerful than humans Lower cost than human power Dung may be used for cooking fuel	Require feeding all year round May be diverted to other activities at crucial irrigation periods
Hydraulic pumps (e.g. rams)	Unattended operation No fuel costs Easy to maintain Low cost Long life High reliability	Require specific site conditions Low output
Wind pumps	Unattended operation Easy maintenance Long life Suited to local manufacture No fuel requirements	Water storage is required for low wind periods High system design and project planning needs Not easy to install
Solar PV	Unattended operation No fuel costs Low maintenance Easy installation Long life (20 years)	High capital costs Water storage is required for cloudy periods Repairs often require skilled technicians
Diesel and gasoline pumps	Quick and easy to install Low capital costs Widely used Can be portable	Fuel supplies may be erratic and expensive High maintenance costs Short life expectancy Noise and fume pollution

demand, leaving the tank empty at the end of the day. This is due to a mismatch between the sizing, pump capacity and the demand profile during the day.

The main applications for solar water pumping are for livestock watering in the USA and Australia. In Africa the systems are used for village water systems and livestock watering, while applications of solar water pumping for irrigation are on the increase especially in India and China.

A solar irrigation system (Figure 3.2) needs to take account of the fact that demand for irrigation water will vary throughout the year. Peak demand during the irrigation seasons is often more than twice the average demand. This means that solar pumps for irrigation can be underused for most of the year although there can be a reduction in strength of the sun during these times, reducing the supply side of the equation. Attention should be paid to the system of water distribution and application to the crops. The system should minimize water losses, without imposing significant additional head on the pumping system, and be of low cost.

20 HANDBOOK OF SMALL-SCALE ENERGY TECHNOLOGIES

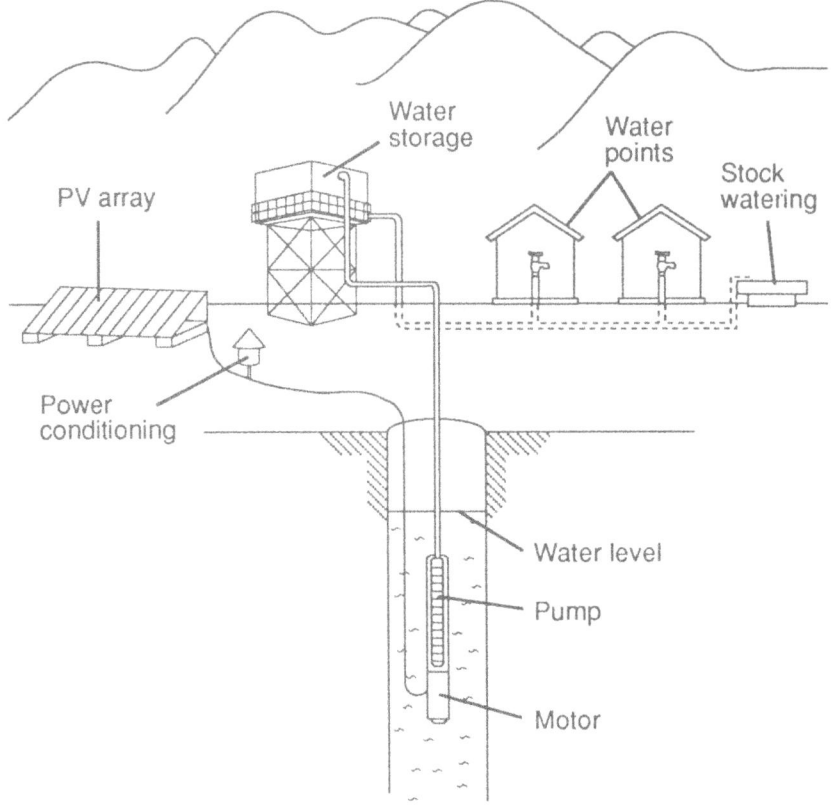

Figure 3.1 Village water supply with solar pumping
Source: Practical Action

The suitability of major irrigation systems for use with solar pumps is shown in Table 3.2.

Table 3.2 Suitability of irrigation methods for use with solar pumps

Irrigation method	Typical application efficiency (%)	Typical head (m)	Suitable for use with solar pumps
Open channels	50–60	0.5–1	Yes
Sprinkler	70	10–20	No
Trickle/drip	85	1–2	Yes
Flood	40–50	0.5	No

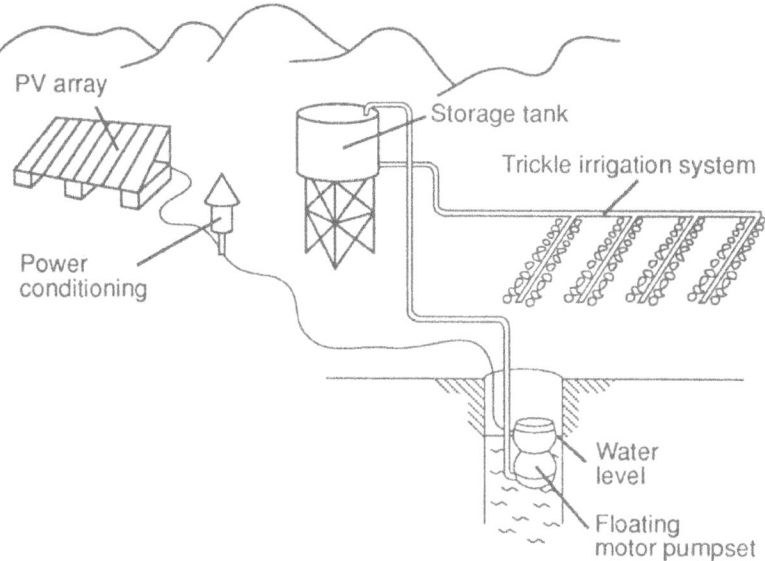

Figure 3.2 Solar irrigation system
Source: Practical Action

Components of solar pump systems

Photovoltaic pumps are made up of a number of components. There is a photovoltaic array which converts solar energy directly into electricity as DC. The pump will have an electric motor to drive it. The characteristics of these components need to be matched to get the best performance. The pump motor unit will have its own optimum speed and load depending on the type and size of the pump.

Motor

This can be DC or the more widely available AC. If an AC motor is used then an inverter is also needed. Inverters have become cheap and efficient and solar pumping systems use special electronically controlled variable-frequency inverters which will optimize matching between the panel and the pump. A typical AC system would also need batteries which require maintenance and add to the cost as the system is less efficient and necessitates a larger array.

The most efficient type of DC motor is a permanent magnet motor. DC motors may have carbon brushes which need replacing when they wear out. If a brushed DC motor is used then the equipment will need to be pulled up from the well (approximately every two years) to replace brushes. Brushless designs of DC motors exist where electrical circuits are used instead of commutators and brushes. These are becoming popular in solar pumping systems. Brushless DC motors would require electronic commutation.

Solar panels

The basic principles of solar photovoltaic panels are explained in Chapter 2, 'Solar photovoltaic energy' (Noble, 2012). Some models use a GPS sensor to provide latitude, longitude and time data to enable the controller to calculate the position of the sun and position the solar array.

The pump

Pump options and the system configuration are described here.

Submersible pumps. Often with electronic load controllers, the pump will be submerged while the load controller is above ground. The advantages of this configuration are that it is easy to install, often with lay-flat flexible pipework, and the motor pumpset is submerged away from potential damage.

Multistage centrifugal pumps. The centrifugal pump will start at low torque and can be matched with the solar array without electronic controllers. The pump is not as efficient as a positive displacement pump using a cheap electronic load controller. Centrifugal pumps are suitable for smaller heads; the older type of set with AC motors can operate at heads of 10–25 m.

Positive displacement helical pumps. The helical pump has the best efficiency and the smallest PV panel for the same specifications of water delivery volume, pressure and head. It has low rotational speed. The pump is made up of a metal helical rotor which rotates in a rubber casing, and is suitable for bigger heads.

A mono solar pump will slow down when the weather is cloudy, but because it has no minimum speed (unlike a centrifugal pump) it will keep delivering water.

Submerged pump with surface-mounted motor. The main advantage is easy access to the motor for maintenance. The low efficiency from power losses in the shaft bearings and the high cost of installation have been its main disadvantages. In general this configuration is largely being replaced by the submersible motor and pumpset.

Floating motor pumpsets. The versatility of the floating unit set makes it ideal for irrigation pumping for canals and open wells. The pumpset is easily portable and there is a negligible chance of the pump running dry. Most of these types use a single-stage submerged centrifugal pump. The most common type has a brushless DC motor. Often the solar array support incorporates a handle or 'wheelbarrow' type trolley to enable transportation.

Surface suction pumpsets. This type of pumpset is also suitable for low head applications. It is not recommended except where an operator will always be in attendance for maintenance and security of exposed systems. Although the use of primary chambers and non-return valves can prevent loss of prime, in practice self-start and priming problems are experienced. It is not practical to have suction heads of more than 8 metres.

Less common types of solar-powered pumps include reciprocating piston (nodding donkey) pumps, and solar thermal pumps or thermosyphon pumps exist but are not commercially used.

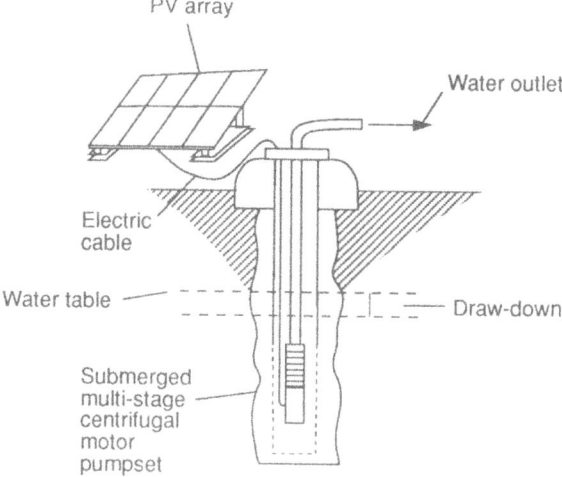

Figure 3.3 Submerged multistage centrifugal motor pumpset
Source: Practical Action

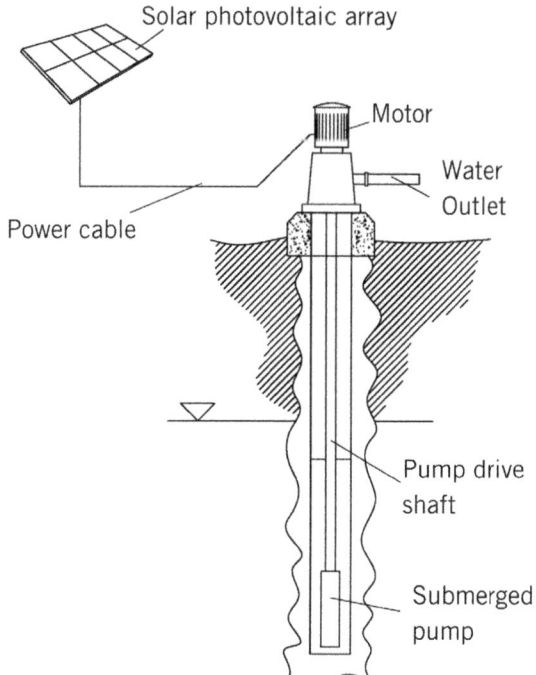

Figure 3.4 Submerged pump with surface-mounted motor
Source: Practical Action

Performance

Solar pumps are available to pump from a head of up to 200 m and with outputs of up to 250 m³/day. Solar pumping technology continues to improve. In the early 1980s the typical solar energy to hydraulic (pumped water) energy efficiency was around 2 per cent, with the photovoltaic array being 6–8 per cent efficient and the motor pumpset typically 25 per cent efficient. Today, an efficient solar pump might have an average daily solar energy to hydraulic efficiency of more than 9 per cent but lower efficiencies of 2–3 per cent are still common. It is important to get the most efficient pump available as the difference in cost between the poor pump and a very efficient pump is much less than the additional cost required for a larger PV panel. Accurate sizing of the array is important in keeping costs down.

A good subsystem (i.e., the motor, pump and any power conditioning equipment used to help maintain a smooth voltage) should have an electrical to hydraulic efficiency of around 70 per cent using positive displacement pumps. With diaphragm pumps the efficiency will be around 45 per cent and centrifugal pumps might have an efficiency of 20 per cent.

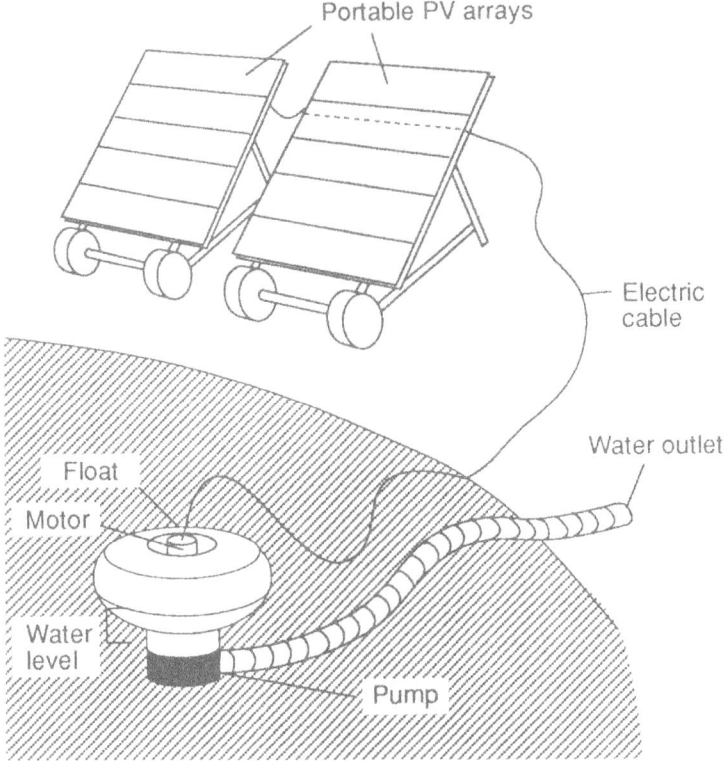

Figure 3.5 Floating motor pump

Source: Practical Action

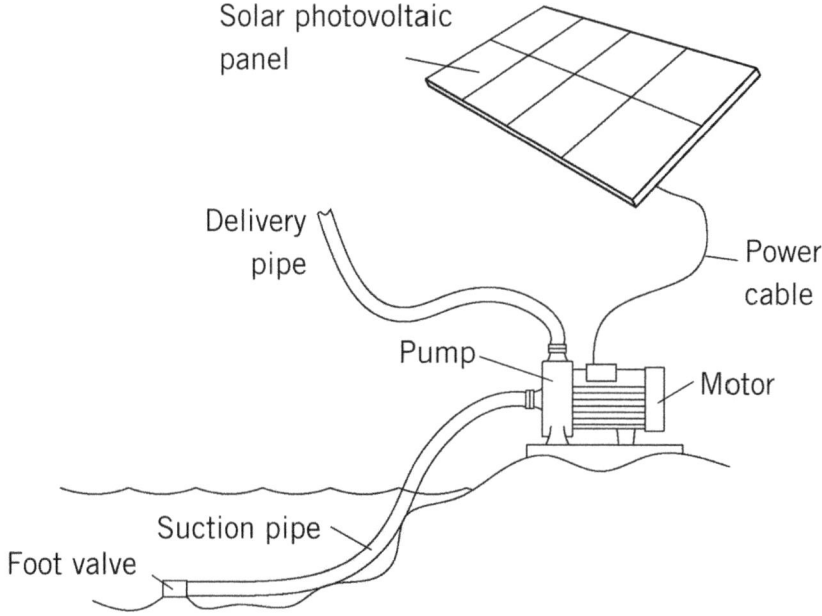

Figure 3.6 Suction pumpsets
Source: Practical Action

Procurement – assessing requirements

The output of a solar pumping system is very dependent on good system design derived from accurate site and demand data. It is therefore essential that accurate assumptions are made regarding water demand/pattern of use and water availability including well yield and expected drawdown.

Domestic water use per capita tends to vary greatly depending on availability. The long-term aim is to provide people with water in sufficient quantities to meet all requirements for drinking, washing and sanitation. WHO guidelines suggest a per capita provision of 40 to 50 litres per day for domestic use only, thus a village of 500 people has a requirement of 20 m³/day. Most villages have a need for combined domestic and livestock watering which will necessitate much greater amounts of water.

Irrigation needs depend upon crop water requirements, effective groundwater contributions and efficiency of the distribution and field application system. Irrigation requirements can be determined by consultation with local experts and agronomists or by reference to technical documents (Allen et al., 1988; Cornish, 1998; Stern, 1988, 1979; and Practical Action Technical Brief on Micro-Irrigation).

Assessing water availability

Several water source parameters need to be taken into account and where possible measured. These are:

- the depth of the water source below ground level;
- the height of the storage tank or water outlet point above ground level;
- seasonal variations in water level.

The drawdown or drop in water level after pumping has commenced also needs to be considered for well and borehole supplies. This will depend on the ratio between pumping rate and the rate of refill of the water source, and should be measured and/or provided by those who drill the borehole. In addition, there is usually a seasonal variation in the water level, and a long-term trend in the water table level dropping.

The pattern of water use should also be considered in relation to system design and storage requirements. Water supply systems should include sufficient covered water storage to provide for daily water requirements and short periods of cloudy weather. Generally, two to five days' water demand is stored.

Sizing solar pumps

The hydraulic energy required (kWh/day)

$$= \text{volume required (m}^3\text{/day)} \times \text{head (m)} \times \text{water density} \times \text{gravity} / (3.6 \times 10^6)$$

$$= 0.002725 \times \text{volume (m}^3\text{/day)} \times \text{head (m)}$$

The solar array power required (kWp)

$$= \frac{\text{hydraulic energy required (kWh/day)}}{\text{av. daily solar irradiation (kWh/m}^2\text{/day} \times F \times E)}$$

where F = array mismatch factor = 0.80 on average (a safety factor for real panel performance in hot sun and after 10–20 years)

and E = daily subsystem efficiency = 0.25–0.40 typically

Economics

In general photovoltaic pumps are economic compared to diesel pumps up to approximately 3 kWp for village water supply and to around 1 kWp for irrigation.

Further information

Technical Briefs and other fact sheets

A Cost and Reliability Comparison Between Solar and Diesel Powered Pumps, Solar Electric Light Fund (SELF) (2008), www.self.org/SELF_White_Paper_-_Solar_vs_Diesel.pdf

Micro-Irrigation, http://practicalaction.org/micro-irrigation

Theft Prevention Measures for PV Systems, Emcon Consulting Group, Namibia, www.emcongroup.com/downloads/Theft%20prevention%20measure%20for%20solar%20PV%20installations.pdf

Bibliography

Allen, Richard G., Pereira, Luis S., Raes, Dirk and Smith, Martin (1988) *Crop Evapotranspiration – Guidelines for Computing Crop Water Requirements,* Irrigation and drainage paper 56, FAO, Rome. Available at: www2.webng.com/bahirdarab/Evapotranspiration.pdf

Cornish, Gez (1998) *Modern Irrigation Technologies for Smallholders in Developing Countries*, Practical Action Publishing, Rugby.

Fraenkel, Peter and Thake, Jeremy (2006) *Water Lifting Devices: A Handbook,* 3rd edn, Practical Action Publishing, Rugby.

Green Empowerment (2007) *Solar Pumping Systems: Introductory and Feasibility Guide*, Green Empowerment, Portland, OR. Available at: www.greenempowerment.org/attachments/SolarWaterPumpingManual_jan2007.pdf

Kenna, Jeff and Gillett, Bill (1985) *Solar Water Pumping: A Handbook*, Practical Action Publishing, Rugby.

Louineau, Jean-Paul (2008) *Practical Guide to Solar Photovoltaic Systems for Technicians: Sizing, Installation and Maintenance,* Practical Action Publishing, Rugby.

Sanchez, Teodoro (2005) *Electricity Services in Remote Rural Communities: The Small Enterprise Model,* Practical Action Publishing, Rugby.

Noble, Neil, ed. (2012) *A Handbook of Small-scale Energy Technologies: Practical Answers*, Practical Action Publishing, Rugby, UK.

Stern, Peter (1979) *Small-scale Irrigation,* Practical Action Publishing, Rugby.

Stern, Peter (1988) *Operation and Maintenance of Small Irrigation Schemes,* Practical Action Publishing, Rugby.

Yaron, Gil, Forbes, Tani and Jansson, Sven (1994) *Solar Energy for Rural Communities: The Case of Namibia,* Practical Action Publishing, Rugby.

Useful addresses

Green Empowerment
140 SW Yamhill, Portland, OR 97204, USA
Tel: +1 503 284 5774, Web: www.greenempowerment.org

Lifewater International
2840 Main Street, Morro Bay, CA 93442
Tel: +1 805 772 0600, Web: www.lifewater.org/technical-library

Suppliers

The most suitable suppler will depend on the part of the world in which the pump is to be installed. Information on equipment suppliers can be found at http://energy.sourceguides.com/ and www.ecobusinesslinks.com/solar_energy_solar_power_panels.htm. This is a sample list of suppliers; it does not imply endorsement by Practical Action.

African Energy
PO Box 664, Saint David, AZ 85630, USA
Tel: +1 520 720 9475, Web: www.africanenergy.com
A specialized distributor of solar electric and power back-up equipment focusing exclusively on the African market, including solar water pumping systems

Davis and Shirtliff Limited
PO Box 41762-00100, Nairobi, Kenya
Tel: +254 020 6968 000, Web: www.dayliff.com
Solar water pumping equipment suppliers

Grundfos International A/S
Poul Due Jensens Vej 7, Bjerroingbo, DK-8850 Denmark
Tel: +45 86 68 1400, Web: www.grundfos.com
Solar- and wind-powered water pumping hybrids – SQFlex

Mono Pumps (Australia) Pty Ltd
338–48 Lower Dandenong Road, Mordialloc 3195, Victoria, Australia
Web: www.monopumps.com.au/en-au/solar-products
Solar-powered water pumping equipment with helical pump

Tata BP Solar India Ltd
78 Electronics City, Hosur Road, Bangalore 560 100, India
Tel: +91 6660 1300, Web: www.tatabpsolar.com
A wide range of solar products including solar water pumping

Tenesol
Z.A.C. de la Tour, 12–14 allée du Levant, 69890 La Tour de Salvagny, France
Tel: +33 (0)4 78 48 88 50, Web: www.tenesol.com/en

True Energy Group
Pendre Enterprise Park, Tywyn, Gwynedd, Wales, LL36 9LW, UK
Tel: +44 (0) 1654 712 713, Web: www.trueenergy.com
Professional renewable energy products and solutions suppliers. International experience on solar water pumping

Chapter 4

Solar thermal energy

This chapter looks at the various technologies that use heat generated from the sun. It covers the basics of solar radiation and summarizes common solar thermal energy applications, such as water heating, drying and cooking. Solar thermal considerations in architecture and greenhouse applications are discussed. Less common applications, such as solar thermodynamic water pumping and solar washing machines, are also presented.

Keywords: solar, thermal, renewable, energy, water heating, water pumping, greenhouse

Introduction

Although most research into the use of solar energy in recent years has been on photovoltaics, where sunlight is converted directly into electricity, there are many applications of solar thermal energy, such as heating, drying and water distillation. Many solar thermal technologies have existed for centuries and are well understood. Manufacturing bases have been established in many sun-rich countries. This can be done on a small scale without using expensive equipment. More sophisticated solar thermal technologies do exist that generate electricity (often on a large scale) but these are not covered here. Solar technologies that rely entirely on energy absorbed from the sun and have no moving components are referred to as *passive solar* technologies whereas *active solar* technologies may have some additional input such as a pump to drive the system.

The nature and availability of solar radiation

Solar irradiation or insolation is the 'rate of delivery of direct solar radiation per unit of horizontal surface', measured in W/m^2 (merriam-webster.com). The earth revolves around the sun with its axis tilted at an angle of 23.5 degrees. It is this tilt that gives rise to the seasons. The strength of sun is dependent upon the angle at which it strikes the earth's surface and as this angle changes during the year, so the insolation changes. Thus, in northern countries, in the depths of winter, where the sun is low in the sky to the south, the radiation strikes the earth's surface obliquely and solar energy is low.

These two phenomena provide an explanation for the variations of solar irradiation with season and latitude. The total solar irradiation received in a day can vary from 0.5 kWh/m^2/day in the UK winter to 5 kWh/m^2 in the UK summer and can be as high as 7 kWh/m^2/day in desert regions of the world, such as regions of

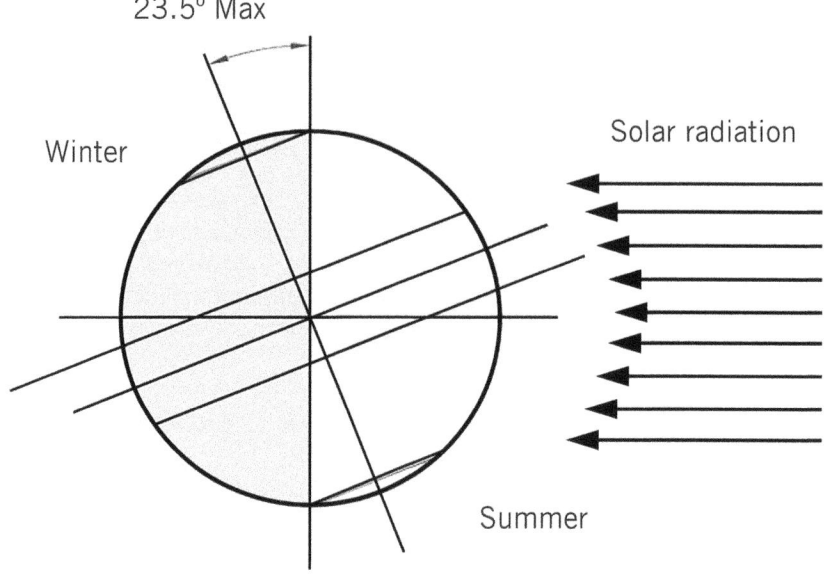

Figure 4.1 The angle of the earth to the sun changes throughout the year

Nigeria (Agbo and Oparaku, 2006) and the Sahara in Algeria (Survey of Energy Resources, 2010). Many tropical regions do not have large seasonal variations and receive an average 6 kWh/m^2/day throughout the year.

Figure 4.2 shows the approximate percentages of direct and diffuse solar insolation that reaches the surface of the earth. As the direct insolation forms a larger proportion of the total received, it follows that varying factors such as the weather, i.e. cloud cover, and the time of day will greatly affect the amount of solar insolation reaching the surface of the earth.

While both direct and diffuse radiation are useful, diffuse radiation cannot be concentrated. Although seemingly obvious, it is also useful to note that most of the solar water heating will occur during daylight hours. This does not necessarily coincide with demand for hot water, so solar hot water storage tanks are normally required.

Daily, seasonal and geographical variations in solar insolation are an important aspect of solar energy because of the influence on system design and solar energy economics. A useful document which summarizes the extent of the application of solar energy in 43 countries around the world is the *2010 Survey of Energy Resources* by the World Energy Council. The document also notes the levels of solar radiation that can generally be expected in the countries listed which may be useful as a quick reference.

Alongside journals and books, another useful source analysing solar insolation levels around the world has been developed by NASA and is free for public use.

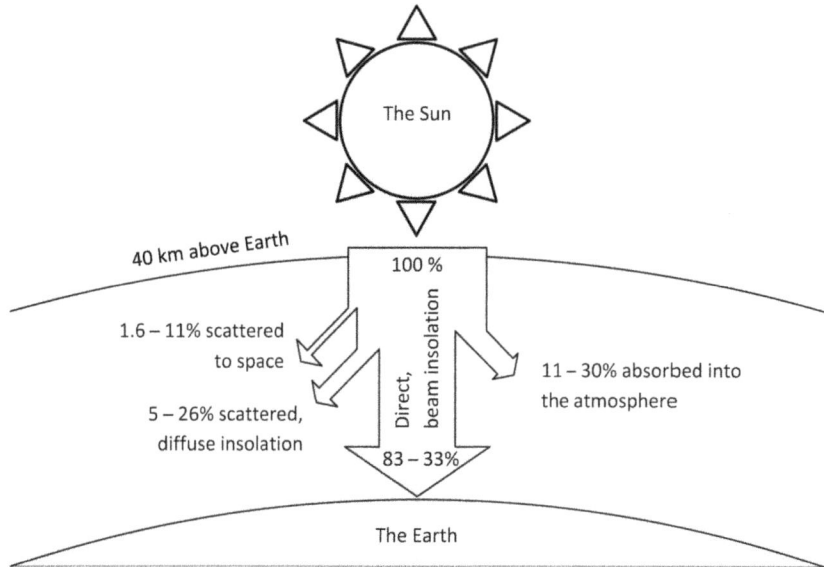

Figure 4.2 Dispersion of solar irradiance through the atmosphere

Source: powerfromthesun.net

The analysis tool enables users to quantitatively investigate the variations in solar insolation in a particular region throughout the year. Three snapshots of either a daily average, average across eight days or monthly average can be compared and presented in several different ways: a probe, which gives the insolation at a particular point on the globe; a transect through a region; or an average across a defined region. All three methods can be specified by the user. Each set of data can also be translated into the Google Earth software. This can be used to gain an overall impression of global solar insolation variation at a particular time.

Solar thermal energy applications

Solar energy reaches the earth's surface as short-wave radiation, is absorbed by the earth and objects on the earth that heat up, and then it is re-radiated as long-wave radiation. Obtaining useful power from solar energy is based on the principle of capturing the short-wave radiation and preventing it from radiating away into the atmosphere. For storage of this trapped heat, a liquid or solid with a high thermal mass is used. In a water heating system this will be the fluid that runs through the collector, whereas in a building the walls will act as the thermal mass. Pools or lakes are sometimes used for seasonal storage of heat.

Glass will allow short-wave radiation to pass through it but prevents long-wave radiation heat escaping. If this energy is being used to heat water with a collector panel, then the tilt and orientation of the panel is critical to the level of energy captured and hence the temperature of the water. The collector surface should be orientated towards the sun as much as is possible. Most solar water-heating collectors are fixed permanently to roofs of buildings and cannot be adjusted. More sophisticated systems for power generation use tracking devices to follow the sun through the sky during the day.

There are many methods available for aiding system design and for predicting the performance of a system. The variability of the solar resource is such that any accurate prediction requires complex analytical techniques. Simpler techniques are available for an approximate analysis.

Water heating

The most common use of solar thermal technology is domestic water heating. Hundreds of thousands of domestic hot water systems are in use throughout the world, especially in areas such as the Mediterranean and Australia where there is high solar insolation (the total energy per unit area received from the sun). Domestic water heaters are usually only found amongst wealthier sections of the community in developing countries.

Low-temperature (below 100°C) water heating is required in most countries of the world for both domestic and commercial use. A wide variety of solar water heaters are available. The simplest is a piece of black plastic pipe, filled with water, and laid in the sun for the water to heat up. Simple solar water heaters usually comprise a series of pipes that are painted black and run through an insulated box fronted with a glass panel, known as a solar collector. The fluid to be heated passes through the collector and into a tank for storage. The fluid can be cycled through the tank several times if necessary to raise its heat to the required temperature. There are two common simple configurations for such a system.

The thermosyphon system makes use of the natural tendency of hot water to rise above cold water. The tank in such a system is always placed above the top of the collector and as water is heated in the collector it rises and is replaced by cold water from the bottom of the tank. This cycle will continue until the temperature of the water in the tank is equal to that of the panel. Where there is a mains water supply, fresh cold water is fed into the system from the mains as hot water is drawn off for use. A one-way valve is usually fitted in the system to prevent the reverse occurring at night when the temperature drops.

Open loop systems allow water to run through the solar panels and be stored in the storage tank ready for use. Closed loop systems are where the water that circulates through the solar panel is separate from the water used. The system uses a heat exchanger. This means that antifreeze can be added to the water running through the panels which allows them to be used in cold climates. Atmospheric systems are used where there is no mains water delivery to the storage tank so as water is taken from the hot water tank it is replaced from an additional cold water tank that is located above it. A break pressure valve allows water to feed the hot water tank when required. Atmospheric systems can be open loop or closed loop. Batch solar water heating systems are used as a simple approach to obtaining hot water. The system is filled with water and left to heat up. Once the water is heated up it can be used as required but the system has to be refilled manually. More than 90 per cent of systems worldwide are based on the thermosyphon principle.

Pumped solar water heaters use a pumping device to drive the water through the collector. The advantage of this system is that the storage tank can be sited below the collector. The disadvantage of course is that electricity is required to drive the pump. Often the fluid circulating in the collector will be treated with an anti-corrosive and/or antifreeze chemical. In this case, a heat exchanger is required to transfer the heat to the consumer's hot water supply. Integrated systems combine the function of tank and collector to reduce cost and size.

Water heating systems can be made relatively simply while more sophisticated systems are available at a higher price. Evacuated tube collectors have the heat absorbing element placed within an evacuated glass sheath to minimize heat losses. System complexity also varies depending on use. For commercial applications, banks of collectors are used to provide larger quantities of hot water as required. Many such systems are in use at hospitals in developing countries.

A solar pond is an approach that uses a large body of water to collect and store solar thermal energy with relatively little equipment. The pond uses the principle that salt water is heavier than fresh water, so a layer of salt water at the bottom of the pond traps the heat energy and the temperature can rise above 90°C. Solar ponds can be used to provide heating to houses and for other applications. For example, using a low-temperature turbine the solar pond can be used to generate electricity or it can be used to provide power to a water distillation unit as developed by the University of Texas (see www.energyeducation.tx.gov/renewables/section_3/topics/solar_ponds/d.html).

Bhuj, Gujarat, has the largest operating solar pond in India, covering an area of 6000 m^2, which is used to supply process heat to the Kutch dairy. The solar pond was developed by the Gujarat Energy Development Agency (GEDA), the Tata Energy Research Institute (TERI) and the Gujarat Dairy Development Corporation (GDDC) (see www.teriin.org/tech_solarponds.php).

Solar drying

Controlled drying is required for various crops and products such as grain, coffee, tobacco, fruits, vegetables and fish. Their quality can be enhanced if the drying is properly carried out. Solar thermal technology can be used to assist with the drying of such products. The main principle of operation is to raise the heat of the product, which is usually held within a compartment or box, while at the same time passing air through the compartment to remove moisture. The flow of air is often promoted using the 'stack' effect which takes advantage of the fact that hot air rises and can therefore be drawn upwards through a chimney, while drawing in cooler air from below. Alternatively a fan can be used. The size and shape of the compartment varies depending on the product and the scale of the drying system. Large systems can use large barns while smaller systems may have a few trays in a small wooden housing.

Coconut drying in Bangladesh
Source: Practical Action / Neil Cooper

Solar crop-drying technologies can help reduce environmental degradation by reducing the use of fuel wood or fossil fuels for crop drying and can also help to reduce the costs associated with these fuels and hence the cost of the product. Helping to improve and protect crops also has beneficial effects on health and nutrition.

Solar wood kilns have been developed to season timber. The solar kiln is constructed using a wooden frame with a glass roof. The principles are similar to that of a greenhouse design. The simplest have been modified greenhouses designed to take large pieces of wood for drying. Air is passed through the kiln to remove the moisture from the wood and the cycle continues.

Solar cooking

Solar cookers fall into two main categories – solar ovens and direct solar concentrators. The basic design for a solar oven is that of a box with a glass cover. The box is lined with insulation and a reflective surface is applied to concentrate the heat onto the pots. The other approach is to reflect the sun's rays onto a pot, often with a parabolic dish. The pots can be painted black to help with heat absorption.

On a domestic scale the cookers have limitations in terms of being effective only during hours of strong sunlight. Another cooking stove is usually required for the periods when there is cloud or during the morning and evening. Cooking time is often a lot slower than conventional stoves and cooking practice has to be adapted to suit. However, the main advantage to solar cookers is that wood does not need to be purchased or collected, which is often a very time-consuming activity for women.

Many variations of solar cooker have been developed from the very basic box made from sheets of reflective cardboard to the very sophisticated large-scale institutional and commercial solar cookers now being used in India (see Practical Action Technical Brief on Solar Cooking).

Desalination and distillation

Basic solar stills can be used to purify water in remote regions where contaminated water is present. They can be used to remove impurities such as fluoride and salts to produce drinking water. The basic still is made of a glass or transparent plastic cover and a shallow tray of water which has a black backing to trap energy. As the sun heats the water within the still, some water evaporates and then condenses on the underside of the covering glass. This glass is at an angle, so the purified water drains off and is captured in a trough separate from the contaminated water. Solar distillation can be combined with other useful functions so that a solar still may also be used for rainwater harvesting if modified slightly. See Chapter 6, 'Solar distillation' (Noble, 2012).

Solar pasteurization

In pasteurization, water is heated to 65°C (149°F) for about six minutes, killing all the germs, viruses, and parasites that cause disease in humans, including cholera and hepatitis A and B. This is similar to what is done with milk and other beverages. It is not necessary to boil the water as many people believe. Pasteurization is not the only way to decontaminate drinking water, but it is particularly easy to scale down so that the initial cost is low (*Boiling Point*, 1995).

Cooling

Preservation of crops and food can be improved with relatively simple evaporative cooling techniques. This approach keeps produce fresh by using the evaporation of water to reduce the temperature and minimize the impact of the sun's energy. Evaporative cooling works best in dry climates; the more humid the air, the less effective it is at reducing the temperature. A similar approach has been used to keep buildings cool by placing a ceramic pot containing water by the window so that the air passes over the pot as it enters the building. This results in a cooler, wetter environment in the room.

Solar thermal energy in architecture

There are two basic requirements within buildings – heating and cooling. Many technological advances have been made in design of 'solar buildings' for solar heating in developed countries but the technology is often expensive and out of reach for rural communities in developing countries.

Space cooling

The majority of the world's developing countries lie within the tropics and most do not need space heating so much as space cooling. There are many traditional, efficient techniques for cooling dwellings, often using effects promoted by the passive solar phenomenon.

There are many methods for minimizing heat gain. These include siting a building in shade or near water, using vegetation or landscaping to direct wind into the building, good town planning to optimize the prevailing wind and available shade. Buildings can be designed for a given climate – for example, domed roofs and thermally massive structures in hot, arid climates; shuttered and shaded windows to prevent heat gain; open-structure bamboo housing in warm, humid areas. In some countries dwellings are constructed underground and take advantage of the relatively low and stable temperature of the surrounding ground.

Space heating

In colder areas of the world (including high altitude areas within the tropics) space heating is often required during the winter months. Vast quantities of energy can be used to achieve this. If buildings are carefully designed to take full advantage of the solar insolation that they receive then much of the heating requirement can be met by solar gain alone. By incorporating certain simple design principles a new dwelling can be made to be fuel efficient and comfortable for habitation. Most of these principles are architecture-based and passive in nature. The use of building materials with a high thermal mass (which stores heat), good insulation and large glazed areas can increase a building's capacity to capture and store heat from the sun. Many technologies exist to assist with daily heating needs but seasonal storage is more difficult and costly.

For passive solar design to be effective, certain guidelines should be followed:

- A building should have large areas of glazing facing the sun to maximize solar gain.
- Features should be included to regulate heat intake to prevent the building from overheating.

- A building should be of sufficient mass to allow heat storage for the required period.
- Features should be incorporated which promote the even distribution of heat throughout the building.

One example of a simple passive space heating technology is the Trombe wall. A massive black-painted wall has a double-glazed skin to prevent captured heat from escaping. The wall is vented to allow the warm air to enter the room at high level and cool air to enter the cavity between the wall and the glazing. Heat stored in the wall during the day is radiated into the room during the night. This type of technology is useful in areas where the nights are cold but the days are warm and sunny.

Greenhouses

It is possible to expand the diversity of crops grown in mountain areas and therefore enhance the nutritional balance of family diets by the use of simple greenhouse structures. However, the low-cost materials such as polythene sheets and wooden pole frames are vulnerable to damage in the harsh climate of mountainous regions.

Sea-water greenhouse

A solar energy application in which evaporative cooling and desalination are combined can create an enhanced environment for crops to grow in hot arid regions and provide clean water. Sea water is fed onto porous cardboard evaporators and humidified air is drawn into the greenhouse. This provides a cooling effect that reduces the temperature within the greenhouse. At the other end of the greenhouse a plastic condenser captures the clean water from the air. The condenser uses cold sea water as a coolant. The greenhouse is made of a light steel structure with a polythene covering. The polythene films are treated to incorporate ultraviolet-reflecting and infrared-absorbing properties. The cardboard evaporators become strengthened by the crystallized calcium carbonate from the sea water.

Less common applications

Solar thermodyamic water pumping

Many solar water pumping systems are based on photovoltaic technology combined with a battery storage system and an electric pump. Solar thermodynamic systems use the heat from the sun to power a pump.

The system can be divided into the following components:

- solar collector which converts radiation to heat;
- heat engine which uses a thermodynamic cycle to convert heat into mechanical energy;
- pump;
- water storage and distribution system.

To convert the thermal energy to mechanical pumping a heat engine is used which can be based on a Rankine cycle or a Stirling cycle. These engines operate through the use of external heat sources including solar energy but any heat source can be used. In addition to driving a water pump, these heat engines can be used in other applications and can generate electricity when combined with a generator.

A solar washing machine

An industrial solar-heated washing machine was developed by FAKT, Energética (a Bolivian NGO), and PROPER (a joint project between GTZ – now GIZ – and the Bolivian Ministry of Energy). The water is heated by solar energy.

The first washing machine was installed in the hospital in Tipuipaya, Bolivia, in 1996 to remove the need to wash by hand. The electricity bill for the hospital accounted for one third of the hospital's budget, so a conventional electric washing machine would have been too costly. This doubled the amount of washing that could be done, but there were difficulties with removing bloodstains which meant that some pre-washing was required.

Other uses

There are many other uses for solar thermal technology, such as solar thermal power stations that generate electricity. These are complex technologies that require large capital investment and are not covered here. Many of the active solar technologies rely on sophisticated, exotic modern materials for their manufacture. This presents problems in developing countries where such materials have to be imported. Some countries do have a manufacturing base for solar thermal products but it is often small and by no means widespread. The market for solar products in developing countries, such as solar water heaters, is small but growing.

Further information

Technical Briefs

Evaporative Cooling, http://practicalaction.org/evaporative-cooling-1
Solar Cooking, http://practicalaction.org/solar-cooking-and-health-1
Solar Water Heating, http://practicalaction.org/solar-water-heating

Bibliography

Agbo, S.G. and Oparaku, O.U. (2006) 'Positive and future prospects of solar water heating in Nigeria', *The Pacific Journal of Science and Technology*, 7(2): 191–8. Available at: www.akamaiuniversity.us/PJST7_2_191.pdf [accessed 20 August 2011].

Boiling Point (1995) 'The solar puddle – a new water pasteurization technique' and 'Simple solar water heater for developing countries', No. 36, GTZ, Practical Action.

Jagadeesh, A. (2000) 'Simple solar water heater for developing countries', *Home Power*, 76: 36–7. Available at: http://files.uniteddiversity.com/Energy/Home.Power.Magazine/Home_Power_Magazine_076.PDF

Merriam-Webster [website], www.merriam-webster.com/dictionary/insolation [accessed 13 September 2001].

Noble, Neil, ed. (2012) *A Handbook of Small-scale Energy Technologies: Practical Answers*, Practical Action Publishing, Rugby, UK.

Rozis, J. and Guinebault, A. (1996) *Solar Heating in Cold Regions*, Practical Action Publishing, Rugby, UK.

Stine, William, Geyer, Michael and Harrigan, R.W. (2001) *Power from the Sun* [website] www.powerfromthesun.net/Book/chapter02/chapter02.html [accessed 20 September 2011].

WEC (2010) *2010 Survey of Energy Resources*, World Energy Council, Solar Energy. Available at: www.worldenergy.org/publications/3040.asp [accessed 5 July 2011].

Useful addresses

NASA Earth Observations, Analysis Tool,
    http://neo.sci.gsfc.nasa.gov [accessed 13 September 2011]
Seawater Greenhouse, www.seawatergreenhouse.com

'Solar thermal energy' was last updated by Amy Punter for Practical Action in November 2011. Amy Punter undertook the revision of the technical brief for Practical Action as a volunteer through the organization's partnership with Engineers Without Borders.

Chapter 5

Solar drying

This chapter describes the basic principles of drying and the operation of solar dryers. It outlines some of the various types of dryer, such as cabinet, tent and tunnel dryers. The difference between sun drying and solar drying is explained, and solar dryers are compared with fuel-fired dryers.

Keywords: solar, thermal, renewable, energy, solar drying, sun drying, food processing

Introduction

The heat from the sun coupled with the wind has been used to dry food crops for preservation for several thousand years. Other crops such as timber need to be dried before they can be used effectively, in building for instance. This sun-drying has often developed into solar drying, where the drying area is enclosed and ventilated – often with polythene, acrylic or glass covering – as a more efficient harnessing of the elements of the drying operation. There are innumerable designs in use and each has its advantages and disadvantages. However, there are three basic designs upon which they are based: solar cabinet dryer, tent dryer, and solar tunnel dryer. These are discussed after a brief description of the principles of drying.

Basic principles of drying

Heat is not the only factor which is important for drying. The rate and efficiency of drying depend on:

- temperature, humidity and quantity of air used;
- size of the pieces being dried;
- physical structure and composition;
- air-flow patterns within the drying system.

The condition, quality and amount of air being passed over and through the pieces to be dried determine the rate of drying. The amount of moisture contained in the air to be used for drying is important and is referred to as absolute humidity. The term relative humidity (RH) is more common and is the absolute humidity divided by the maximum amount of moisture that the air could hold when it is saturated. Relative humidity is expressed as a percentage and fully saturated air would have an RH of 100 per cent. This means that it cannot pick up any more moisture. Air that is saturated at a low temperature will, when heated, have a greater capacity to hold water, so at a higher temperature its RH will be lower. Table 5.1 gives an example of air at 29°C with an RH of 90 per cent. Such air, when heated to 50°C, will then have an RH of only 15 per cent. This means that instead of being able to hold only an extra 0.6 grams of water per kilogram (at 29°C) it is able to hold 24 grams per kilogram. Its capacity to pick up moisture has been increased because it has been heated.

When placed in a current of heated air, food initially loses moisture from the surface; this is the constant rate period. As drying proceeds, moisture is then removed from inside the food material, starting near the outside. Moisture removal becomes more and more difficult as the moisture has to move further from deep inside the food to the surface. This is the falling-rate period. Eventually no more moisture can be removed and the food is in equilibrium with the drying air.

Table 5.1 The effect of air temperature on relative humidity

Air temperature (°C)	RH (%)	Amount of water/kg air needed to reach 100% RH (grams)[1]
29	90	0.6
30	50	7.0
40	28	14.5
50	15	24.0

1 the potential for the air to pick up moisture (RH = relative humidity)

During the falling-rate period, the rate of drying is largely controlled by the chemical composition and structure of the food. Design of a dryer depends on the drying rate curve of the material to be dried but these curves are indicative only and depend on the factors mentioned above. The heat required to evaporate water is 2.26 kJ/kg. Hence, approximately 250 MJ (70 kWh) of energy are required to vaporize 100 kg water. If the ambient air is dry enough, heat input is not essential. The greatest potential for drying crops in a short time is when the ambient air is dry and warm. If the air is warm then less air is needed. This temperature will itself depend mainly upon the air temperature but also on the amount of solar radiation received directly by the food being dried.

Solar drying operation

All dryers need ventilation to be able to dry crops effectively. Air movement can be by natural convection or can be assisted using fans. Solar food drying can be performed successfully in most areas, but how quickly the food dries is affected by the variables indicated above, especially the amount of sunlight and relative humidity. Typical drying times in solar dryers are from one to three days depending on sun, air movement, humidity and the type of food to be dried. Most dryers are black inside, either painted or with black polythene inserts to absorb as much solar radiation as possible.

Cabinet dryers

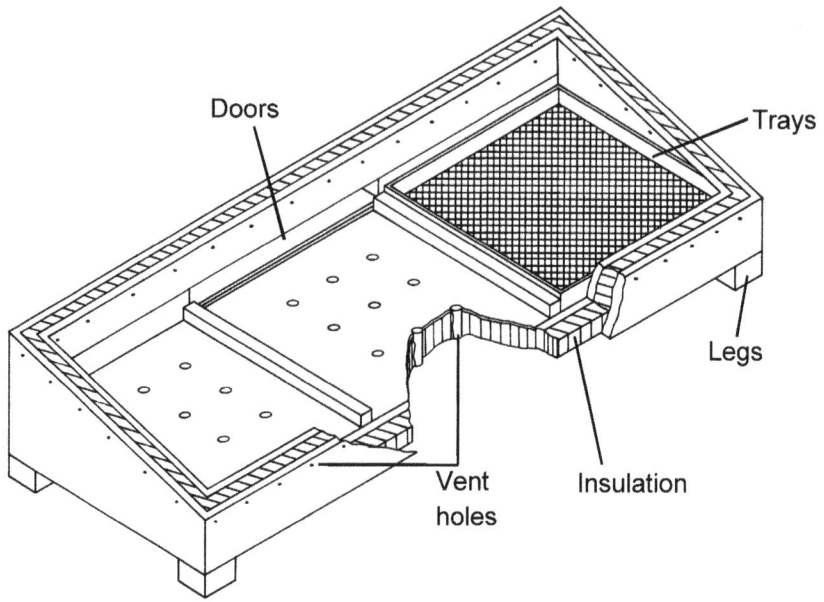

Figure 5.1 The Brace solar cabinet dryer
Source: Practical Action

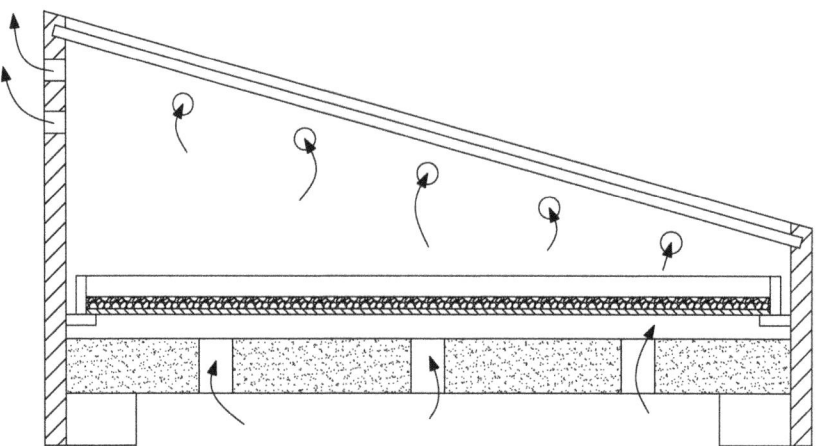

Figure 5.2 Section through the cabinet dryer
Source: Practical Action

Figure 5.2 shows the flow of air in through the vent holes in the underside of the cabinet, past food placed on the drying trays, and out of the holes at the top.

Tent dryers

The distinguishing feature of tent and cabinet dryers is that the drying chamber and the collector are combined (see Figure 5.3). Such dryers provide protection from dust, dirt, rain, wind, and pests. A much smaller tent dryer is shown in the photo. Very similar to a cabinet dryer, it demonstrates the overlap in designs between solar dryers.

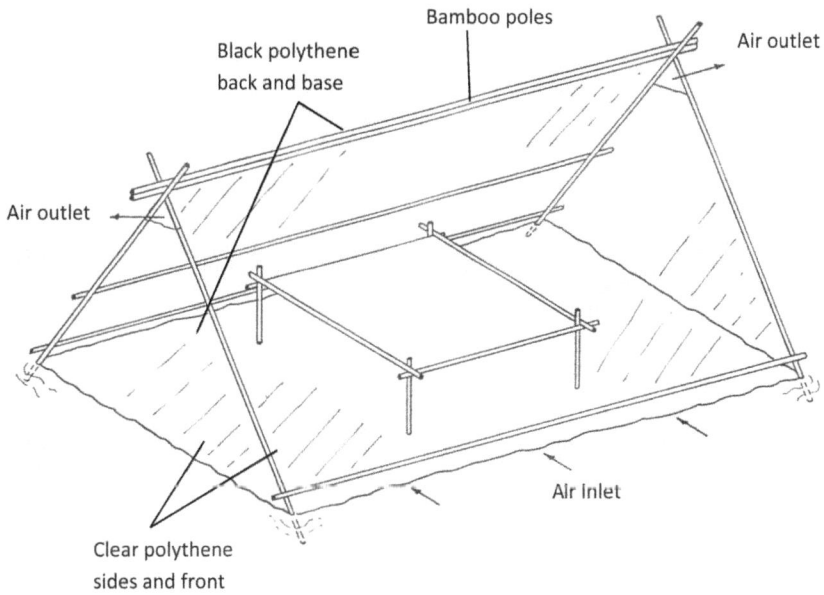

Figure 5.3 Tent dryer

A small solar tent dryer, Ghana

Source: Tony Swetman

Solar tunnel dryers

Many solar dryers employ the use of photovoltaic cells to power fans which blow air across the drying area. Chief among this type of dryer is the Hohenheim dryer produced by Innotech in Germany. By using a fan to create the air flow, operators can substantially reduce drying time. Air flows across the collector area, which is usually painted black to absorb the sun's heat, and is blown across trays containing the material to be dried. Figure 5.4 shows the features of the dryer.

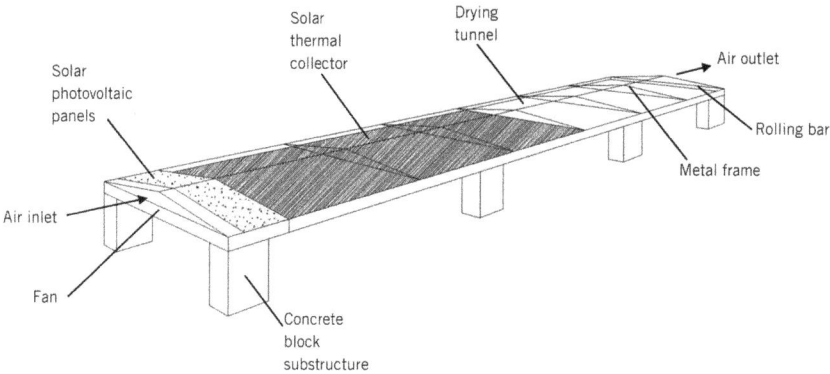

Figure 5.4 Solar tunnel dryer layout

Source: Neil Noble / Practical Action

Black paint being applied to the collector on a Hohenheim dryer, Ghana

Source: Tony Swetman

There are several other types of solar dryer, many involving insulation and different air-flow methods. Some have a chimney fitted to the outlet to encourage better air flow, others make use of external heating sources such as hot water to allow further drying at night or when cloud cover prevents efficient drying. However, all are essentially variations on the three described above.

Sun-drying compared with solar drying

The great advantage of open-air drying is that there is minimal capital outlay. It is labour-intensive, although where labour is cheap this is not a drawback. An important advantage of solar drying is that the product is protected from rain, insects, animals and dust. This improves the hygiene and quality of the product as well as avoiding the need to cover the crop or transfer it to a sheltered area during rain. Solar drying, especially when using fans, gives some control of the drying process at elevated temperatures and can be faster, which reduces the likelihood of mould growth and spoilage of the product. However, care is needed when drying food at a high temperature since too-rapid drying can result in the food becoming dry on the outside and still wet on the inside. This is called 'case-hardening'. This can give a false impression that the food is completely dry. During subsequent storage the trapped moisture will migrate to the outside of the food, raising the humidity and resulting in mould growth and spoilage.

Solar dryers compared to fuel-fired dryers

The choice between using solar radiation and fuel fired dryers using, for instance, wood, charcoal, diesel, gas or electricity depends upon the equipment capital cost, raw material to be dried, operating costs of running the dryer, and the likely price to be obtained for the final dried product. Fuelled heating allows much better control of the drying operation than solar heating and does not depend on the weather. However, it is possible to combine solar drying with a fuel source to reduce fuel costs. Such systems include pre-heating of air by solar energy.

Choice of solar dryer

The choice between types of solar dryer will depend on local requirements including scale of operation as well as the budget available. If smallholder farmers wish to dry crops for their own needs then capital cost may well be the main constraint and so low-cost plastic-covered tent or box dryers may be the most suitable choice. However, commercial farmers with an assured market for their product may consider banks of fan-assisted, glass-covered solar dryers more appropriate for their needs.

Further information

Technical Briefs

Drying of Foods, https://practicalaction.org/drying-of-foods
Small-scale Drying Technologies,
    https://practicalaction.org/small-scale-drying-technologies
Drying of Chillies, https://practicalaction.org/drying-chillies-2
Drying of Apricots, https://practicalaction.org/drying-apricots

Bibliography

Axtell, Barrie (2002) *Drying Food for Profit: A Guide for Small Business*, Practical Action Publishing, Rugby.
Axtell, Barrie (1991) *Try Drying It! Case Studies in the Dissemination of Tray Drying Technology,* Practical Action Publishing, Rugby.
NRI (1996) *Producing Solar Dried Fruit and Vegetables for Small-scale Enterprise Development,* Natural Resources Institute, Chatham Maritime, Kent.
Ogunkoya, A.K., Ukoba, K.O. and Olunlade, B.A. (2011) 'Development of a low-cost solar dryer', *Pacific Journal of Science and Technology* 12 (1). Available at: www.akamaiuniversity.us/PJST12_1_98.pdf
Rozis, Jean François (1997) *Drying Foodstuffs,* Backuys Publishers, Leiden.
Thuillier, Fabrice (2002) *Setting up a Food Drying Business: A Step-by-step Guide*, Practical Action Publishing, Rugby.

Useful addresses

Solar Drying Equipment, FAO – labour-saving technologies,
    http://teca.fao.org/sites/default/files/technology_files/Solar%20drying%20equipment.pdf
Planters Energy Network, http://pen.net.in/

Manufacturers and suppliers

ATR Solar (India)
No.1, RMR Complex, 2nd Floor, North Gate, SS Colony (Opp. Devaki Scans), Madurai – 625 010, Tamil Nadu, India
Tel: +91 452 3025400
http://atrsolar.com/solardryer.htm

IndiaMart
http://dir.indiamart.com/impcat/solar-dryer.html
Lists commercial suppliers of solar dryers
Design and fabrication of different types of solar-assisted dryers to customers' specifications. Supply of tray cabinet dryers, tray tunnel dryer, trolley type tunnel dryer, bin dryer, vertical 'V' port dryers, conveyor continuous dryers and FBD dryers.

Innotech Ingenieursgesellschaft mbH
Weilemer Weg 27, D-71155 Altdorf, Germany
Tel: + 49 (0)7031 / 74 47 41
www.innotech-ing.de/Innotech/english/TT-Data.html
Hohenheim tunnel dryer

NRG Technologists Pvt. Ltd
989/6, GIDC Makarpura, Vadodara – 390010, Gujarat, India
Tel: +91 99047 64068
www.nrgtechnologists.com/
Fish solar dryer, food solar dryer, fruit and vegetable solar dryer, solar dryer for silkworm, solar laundry dryer and spirulina dryer

Steelhacks Industries
Plot No. 525, GIDC Estate, Vithal Udhyognagar – 388 121, Dist. Anand, Gujarat, India
Tel: +91 02692 236156 / 9825049997
www.solarheater.co.in/solar-dryer.html
Solar tunnel dryers, portable solar dryers, solar cabinet dryers

Note: This is a selective list of suppliers and does not imply endorsement by Practical Action.

'Solar drying' was produced by Tony Swetman for Practical Action in November 2007 and last updated in May 2012. Tony Swetman is a freelance consultant on agricultural crop production, processing and marketing. His background is in chemistry, food technology and agricultural economics. He previously worked for the Natural Resources Institute on Food Security and Agricultural Development and has worked in numerous developing countries, both long- and short-term.

Chapter 6

Solar distillation

This chapter outlines how water can be treated using simple equipment that allows contaminated water to evaporate and the clean condensation to be collected. The various equipment configurations and processes are described, including the energy equations and guidance on their application.

Keywords: solar, thermal, renewable, energy, distillation, water treatment, evaporation, drinking water

Introduction

Solar water distillation is a very old technology. An early large-scale solar still was built in 1872 to supply a mining community in Chile with drinking water. The technology has been used for emergency situations including Navy introduction of inflatable stills for lifeboats.

Solar distillation is a relatively simple treatment of water that is brackish (i.e. containing dissolved salts). Distillation is one of many processes that can be used for water purification and can use any heating source. Solar energy is a low-tech option. In this process, water is evaporated using the energy of the sun, then the vapour condenses as pure water. This process removes salts and other impurities. Solar distillation is used to produce drinking water or to produce pure water for lead-acid batteries, laboratories, and hospitals, and in producing commercial products such as rose water. It is recommended that drinking water has 100–1000 mg/l of salt to maintain electrolyte levels and for taste. Some saline water may need to be added to the distilled water for acceptable drinking water.

There are a number of other approaches to desalination, such as photovoltaic-powered reverse osmosis, for which small-scale commercially available equipment is available; solar distillation has to be compared with these options to determine its appropriateness to any situation. If treatment of polluted water is required rather than desalination, slow sand filtration is a low-cost option.

Energy requirements for water distillation

The energy required to evaporate water, called the latent heat of vaporization of water, is 2.26 megajoules per kilogram (MJ/kg). This means that to produce 1 litre (i.e. 1 kg, as the density of water is 1 kg/litre) of pure water by distilling brackish water requires a heat input of 2.26 MJ. This does not allow for the efficiency of the system used which will be less than 100 per cent, or for any recovery of latent heat that is rejected when the water vapour is condensed. It should be noted that, although 2.26 MJ/kg or 2260 kJ/kg is required to evaporate water, to pump a kilogram of water through a 20 m head requires only 0.2 kJ/kg. Distillation is therefore normally considered only where there is no local source of fresh water that can be easily pumped or lifted.

How a simple solar still works

The main features are the same for all solar stills. The solar radiation is transmitted through the glass or plastic cover and captured by a black surface at the bottom of the still. A shallow layer of water absorbs the heat and then produces vapour within the chamber of the still. This layer of water should be 20 mm deep for best performance. The vapour condenses on the cover, which is at a lower temperature because it is in contact with the ambient air, and the condensate runs down into a gutter from where it is fed to a storage tank.

Design objectives for an efficient solar still

For high efficiency the solar still should maintain:

- a high feed (undistilled) water temperature;
- a large temperature difference between feed water and condensing surface;
- low vapour leakage.

A high feed water temperature can be achieved if:

- a high proportion of incoming radiation is absorbed by the feed water as heat – hence low-absorption glazing and a good radiation-absorbing surface are required;
- heat losses from the floor and walls are kept low;
- the water is shallow so there is not so much to heat.

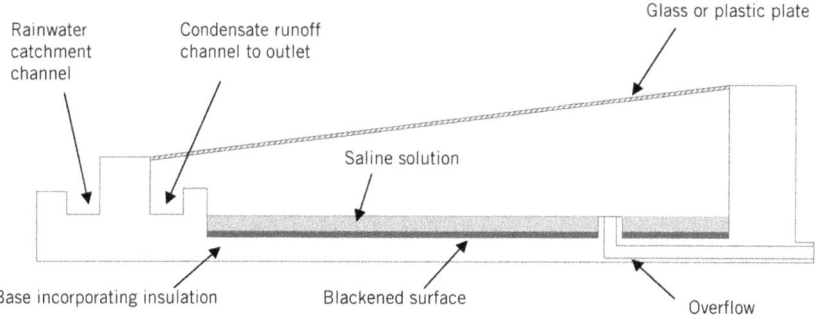

Figure 6.1 A single-basin still

Source: Martin Bounds / Practical Action

A large temperature difference can be achieved if:

- the condensing surface absorbs little or none of the incoming radiation;
- condensing water dissipates heat which must be removed rapidly from the condensing surface by, for example, a second flow of water or air, or by condensing at night.

Technical aspects

Single-basin stills have been much studied and their behaviour is well understood. The efficiency of solar stills which are well-constructed and -maintained is about 50 per cent although typical efficiencies can be 25 per cent. Daily output as a function of solar irradiation is greatest in the early evening when the feed water is still hot but when outside temperatures are falling. At very high air temperatures, such as over 45°C, the plate can become too warm and condensation on it can become problematic, leading to loss of efficiency.

Some problems with solar stills which would reduce their efficiency include:

- poor fitting and joints, increasing colder air flow from outside into the still;
- cracking, breakage or scratches on the glass, reducing solar transmission or letting in air;
- growth of algae and deposition of dust, bird droppings, etc., which must be avoided by cleaning the still regularly every few days.

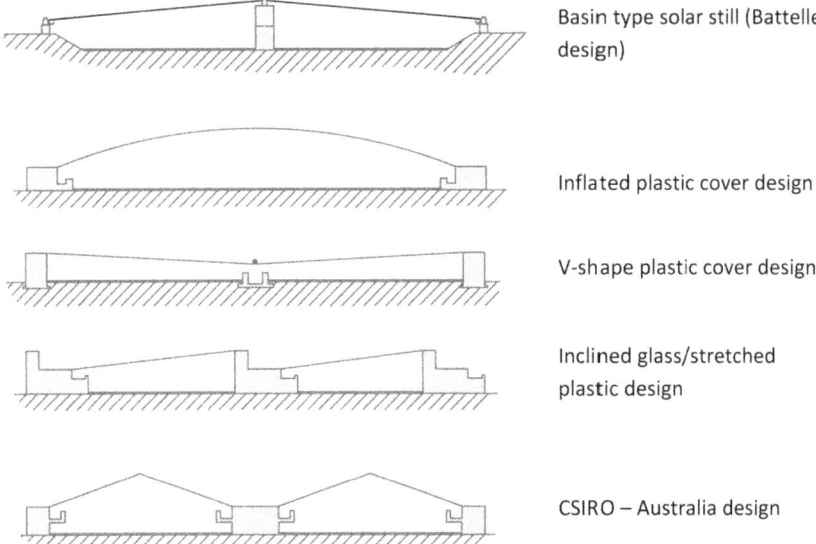

Figure 6.2 Examples of solar still designs

Source: Martin Bounds / Practical Action

SOLAR DISTILLATION   51

- Damage is inflicted over time to the blackened absorbing surface.
- Salt accumulates on the bottom and needs to be removed periodically.
- The saline water in the still is too deep, or dries out. The depth needs to be maintained at around 20 mm.

The cover can be either glass or plastic. Glass is preferable to plastic because most plastic degrades in the long term due to ultraviolet light from sunlight and because it is more difficult for water to condense onto it. Tempered low-iron glass is the best material to use because it is highly transparent and not easily damaged (Scharl and Harrs, 1993). However, if this is too expensive or unavailable, normal window glass can be used. This has to be 4 mm thick or more to reduce breakages. Plastic (such as polythene) is acceptable for short-term use.

A very low angle cover plate at the back reflects sunlight and thus reduces efficiency. Stills with a single sloping cover with the back made from an insulating material avoid this problem. It is important for greater efficiency that the water condenses on the plate as a film rather than as droplets, which tend to drop back into the saline water. For this reason the plate is set at an angle of 10–20 degrees. The condensate film is then likely to run down the plate and into the runoff channel.

Brick, sand, concrete or waterproofed concrete can be used for the basin of a long-life still if it is to be manufactured on-site, but for factory-manufactured stills, prefabricated ferro-concrete can be used. Moulding of stills from fibreglass was tried in Botswana (Yates et al., 1990) and proved to be more expensive than a brick still and more difficult to insulate sufficiently, but with the advantage of the stills being transportable.

By placing a fan in the still it is possible to increase evaporation rates. However, the increase is not large and there is the extra cost and complication of including and powering a fan in what is essentially quite a simple piece of equipment. Fan-assisted solar desalination would only really be useful if a particular level of output is needed when the area available for the stills is restricted, as fan assistance can enable the area occupied by a still to be reduced for a given output.

The Mexican still

In the Mexican still, two stills as described above are fixed together to form a triangular tent shape. The glass plates can be supported from below at the apex where they join, but, if they are not fixed with sealant and just lean against each other, this increases the fragility of the still and limits the area even further of each of the glass plates.

The Brace Research Institute still

This is essentially a still as shown in Figure 6.1. However, several stills are placed next to each other over the width of say 10 metres of the distillation plant. Lengthwise, the unit is built over a considerable distance, such as 15 m. Glass plates are placed along the length of the still and simply joined with sealant. Units of this size also have two small weirs lengthwise to encourage saline water to flow along the full length of the still. A project of this type was set up in Haiti by the Brace Research Institute, McGill University, Canada. The scale of the unit requires caretakers to be trained in its quite considerable maintenance.

Multiple-effect basin stills

These have two or more compartments. The condensing surface of the lower compartment is the floor of the upper compartment. The heat given off by the condensing vapour provides energy to vaporize the feed water above. Efficiency is therefore greater than for a single-basin still, typically being 35 per cent or more, but the cost and complexity are correspondingly higher.

Wick stills

In a wick still, the feed water flows slowly through a porous, radiation-absorbing pad (the wick). Two advantages are claimed over basin stills. First, the wick can be tilted so that the feed water presents a better angle to the sun (reducing reflection and presenting a large effective area). Second, less feed water is in the still at any time and so the water is heated more quickly and to a higher temperature. Simple wick stills are more efficient than basin stills and some designs are claimed to cost less than a basin still of the same output. Some designs have been developed which incorporate absorbent or film-type materials to increase the surface area of evaporation (for example, Tiwari, 1993).

Use of reflector

The inside walls of the still can incorporate a reflective coating, such as aluminium foil, to increase the reflection of heat energy onto the evaporating water. It is not known how far this has helped to improve the efficiency of the still.

Inverted absorber solar stills

Heat is absorbed from the underside of the still to improve efficiency. This allows the condenser plate and the collector plate to be separate, allowing the collector plate to operate at a much higher temperature than the condenser plate. There are several designs of inverted absorber from the fairly simple to more complex designs.

Spherical still. In a design developed by the Thermal and Solar Laboratory at Claude Bernard University, Lyons, France, a trough containing the saline water is positioned in the centre of a hollow transparent plastic sphere. Water condenses on the inside surface of the sphere and is collected by a mechanical windscreen-wiper-type blade which forces it to fall to the bottom of the sphere to be collected. There seems to be a small improvement in efficiency compared

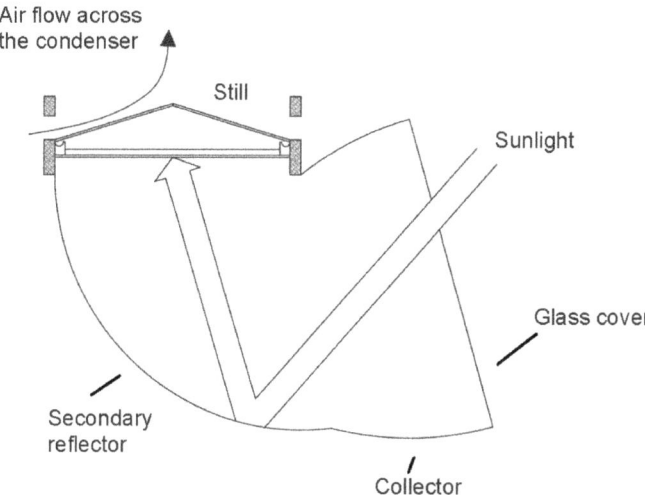

Figure 6.3 An inverted absorber solar still

Source: Neil Noble / Practical Action

with a conventional solar still, but the greater cost of this still might cancel out any advantage.

Inclined stills. The aim of inclining a still is to increase the solar radiation by catching it directly, rather than at an angle as with stills which lie flat. To do this constantly, as the sun rises and sets, would need someone to monitor the sun and turn the unit regularly, or a sophisticated automatic tracking and turning mechanism.

Condensate heat recovery. Heat recovery from the energy given out when water vapour condenses has generally not been attempted with small-scale solar distillation, unlike with larger-scale systems. It is known that the Ben Gurion Institute, and latterly the Technion Institute in Israel, has undertaken some experiments with heat recovery. In the simplest system, saline water is made to flow over the outside of the condensation plate before entering the still, but then this would reduce the amount of solar radiation passing through the plate. There may be scope for further research to overcome current difficulties with attempting heat recovery from solar distillation.

Emergency still. To provide emergency drinking water on land, a very simple still can be made that makes use of the moisture within the earth. All that is required is a plastic cover, a bowl or bucket, and a pebble. Other designs of simple still have been developed for emergencies at sea.

Hybrid designs. Solar stills can usefully be combined with another function of technology. Three examples are:

- *Rainwater collection.* By adding an external gutter, the still cover can be used for rainwater collection to supplement the solar still output.

- *Greenhouse solar still*. The roof of a greenhouse can be used as the cover of a still.
- *Supplementary heating*. Waste heat from an engine or the condenser of a refrigerator can be used as an additional energy input.

Output of a solar still

An approximate method of estimating the output of a solar still is given by:

$$Q = \frac{E \times G \times A}{L}$$

where:

Q = daily output of distilled water (litres/day)
E = overall efficiency
G = daily global solar irradiation (MJ/m²)
L = the latent heat of vaporization of water = 2.26 MJ/kg
A = aperture area of the still, i.e. the plan area for a simple basin still (m²)

The average daily global solar irradiation is typically 18 MJ/m² (5 kWh/m²). A simple basin still operates at an overall efficiency of about 30 per cent. Hence the output per square metre of area is:

$$\text{daily output} = \frac{0.30 \times 18.0 \times 1}{2.26}$$

= 2.39 litres (per square metre)

Performance varies between tropical locations, but not significantly. An average output of 2.3–3 litres/m² per day is typical; the yearly output of a solar still is often therefore referred to as approximately one cubic metre per square metre (1 m³/m²) per year.

Experience

Despite a proliferation of more sophisticated designs such as TERI's solar desalination unit with offset collectors, the single-basin still has the best track record in the field. Hundreds of smaller stills are operating in Africa and India.

The cost of pure water produced depends on:

- the cost of making the still;
- the cost of the land;

SOLAR DISTILLATION

- the life of the still;
- operating costs;
- cost of the feed water;
- the discount rate adopted;
- the amount of water produced.

An example of the cost of a solar still in India is Rs 28,000 for 15 m² (approximately US$575/£368 for 15 m², or $38.30/£24.50 per m², at a conversion rate of $1.00 = £0.64). The price of land will normally be a small proportion of this in rural areas, but may be prohibitive in towns and cities (see the TNAU website for details).

The life of a glass still is usually taken as 20–30 years but operating costs can be high, especially to replace broken glass. It is important that stills are regularly inspected and maintained to retain their efficiency and reduce deterioration. Damage, such as breakage of the collector plate, needs to be rectified.

Some companies – for example, in the United States, Russia, India and South Africa – sell solar stills largely for household use, which can produce up to about 50 litres per day.

Would a solar still suit your needs?

Each person needs 1–2 litres of drinking water a day to live. The minimum requirement for normal life in developing countries (which includes cooking, cleaning and washing clothes) is 20 litres per day (in the industrialized countries 200–400 litres per day is typical). Yet some functions can be performed with salty water and a typical requirement for distilled water is 5 litres per person per day. Therefore 2 m² of still are needed for each person.

Solar stills should normally only be considered for removing dissolved salts from water. If there is a choice between brackish groundwater or polluted surface water, it will usually be cheaper to use a slow sand filter or other treatment device. If there is no fresh water then the main alternatives are desalination, transportation and rainwater collection. Unlike other techniques of desalination, solar stills are more attractive, the smaller the required output. The initial capital cost of stills is roughly proportional to capacity, whereas other methods have significant economies of scale. For the individual household, therefore, the solar still is most economic.

For outputs of 1 m³/day or more, reverse osmosis or electrodialysis should be considered as an alternative to solar stills. Much will depend on the availability and price of electrical power. For outputs of 200 m³/day or more, vapour compression or flash evaporation will normally be the least-cost options. The latter technology can have part of its energy requirement met by solar water heaters.

In many parts of the world, fresh water is transported from another region or location by boat, train, truck or pipeline. The cost of water transported by vehicles is typically in the same order of magnitude as that produced by solar stills. A pipeline may be less expensive for very large quantities.

Rainwater collection is an even simpler technique than solar distillation and is preferable in areas with 400 mm of rain annually, but requires a greater area and usually a larger storage tank. If ready-made collection surfaces exist (such as house roofs) these may provide a less expensive source for obtaining clean water. See the Practical Action Technical Brief on Rainwater Harvesting.

Which solar still?

The single-basin still is the only design proven in the field. Multi-effect stills have the potential to be more economic but it would be as well to gain experience first with a single-basin still.

Further information

Technical Briefs and other fact sheets

Rainwater Harvesting, http://practicalaction.org/rainwater-harvesting-answers
How to Construct a Solar Water Distiller, Practical Action South Asia, http://janathakshan.com/PDFs/SolarwaterDistiller%5Bpdf.pdf

Bibliography

Bloemer, J.W. (1964) *Design of a Basin-Type Solar Still*, UNT Digital Library, Washington DC. Available at: http://digital.library.unt.edu/ark:/67531/metadc11661/m1/9/

Coghlan, A. (1993) 'Technology: Sun-soaked wicks purge salty water', *New Scientist*, 16 January, No. 1856.

Malik, M.A.S., Tiwari, G.N., Kumar, A. and Sodha, M.S. (1982) *Solar Distillation*, Pergamon Press, UK.

Tiwari, G.N. and Tiwari, A.K. (2008) *Solar Distillation Practice for Water Desalination Systems,* Anshan Publishing, Tunbridge Wells, UK.

Yates, R., Woto, T. and Tlhage, J.T. (1990) *Solar-powered Desalination: A Case Study from Botswana*, IDRC, Ottawa.

Useful addresses

Agricultural Engineering College and Research Institute
Tamil Nadu Agricultural University – TNAU, Coimbatore 641 003, Tamil Nadu, India
Web: www.tnau.ac.in/tech/implements/bioen1c.htm
Developed a 15 m^2 solar still

The Centre for Energy Studies, Indian Institute of Technology Delhi – IITD
Hauz Khas, New Delhi 110 016, India
Fax: +91 0112658 2037, Web: www.iitd.ac.in
Researches multi-effect solar distillation systems

El Paso Solar Energy Association (EPSEA)
PO Box 26384, El Paso, TX 79926, USA
Web: www.epsea.org
The website includes details of the projects on installation of solar stills in Texas and Mexico. Construction details of solar stills can be ordered through the website.

Indian Desalination Association (InDA)
Desalination Division, BARC, Trombay, Mumbai 400 085, India
Tel: +91 22 2559 4647, Web: www.indaindia.org/about.htm
An affiliate of IDA

International Desalination Association (IDA)
PO Box 387, 94 Central Street, Suite 200, Topsfield, MA 01983, USA
Tel: +1-978 887 0410, Web: www.idadesal.org
The goals of the association are the development and promotion of the appropriate use of desalination and desalination technology worldwide.

The International Development Research Centre – IDRC
PO Box 8500, Ottawa, ON, K1G 3H9, Canada
Tel: +1 613 236 6163, Web: www.idrc.ca
Project work has included solar distillation in Botswana:
<http://archive.idrc.ca/library/document/053715/chap3_e.html>

Pakistan Desalination Association (PakDA)
42, Bhayani Centre, Block-M, North Nazimabad, Karachi, Pakistan
Tel: +92 21 6677341-2
An affiliate of IDA

The Energy and Resources Institute – TERI
Darbari Seth Block, IHC Complex, Lodhi Road, New Delhi 110 003, India
Tel: +91 11 2468 2100, Web: www.teriin.org/index.php
Design of a multistage solar desalination unit with offset collectors:
<www.teriin.org/index.php?option=com_content&task=view&id=62>

'Solar distillation' was revised by Neil Noble.

Chapter 7
Micro hydropower

Micro hydropower covers wattages from 5kW to 100kW. This brief describes the components of a typical hydro scheme and the suitable conditions for micro hydro. It outlines the energy that can be generated and how to make best use of this energy, introducing the concept of the load factor. The equipment used, such as the types of turbine and load control governor, are described and approaches to minimizing the installation cost on the generator and distribution sides are highlighted.

Keywords: hydro, renewable, energy, turbine, micro hydro, water power, hydroelectricity, electricity generation

Introduction

The most common way to harness water power is by using a turbine which is turned by water moving in a controlled manner. It is a technology that has been used throughout the world, by a diverse range of societies and cultures, for many centuries. Large dams hold water which can be used to provide energy for industry and grid electrification systems. Smaller systems can provide energy to remote regions without the need to build dams.

Table 7.1 outlines the categories used to define the power output from hydropower. Micro hydropower is the small-scale harnessing of energy from falling water; for example, harnessing enough water from a local river to power a small factory or village. This chapter concentrates on micro hydropower.

Water-powered mills have been in use for nearly a thousand years. In Europe, Asia and parts of Africa, waterwheels were used to drive industrial machinery, such as mills and pumps. The first effective water turbines appeared in the mid-nineteenth century and these quickly replaced the older waterwheels in many applications. In contrast to waterwheels and the early turbines, modern turbines are compact, highly efficient and capable of turning at very high speed. Hydropower is a well-proven technology, relying on a non-polluting, renewable and indigenous resource, which can integrate easily with irrigation and water supply projects. China alone has more than 85,000 small-scale, electricity-producing hydropower plants.

Over the past few decades, there has been a growing realization in developing countries that micro-hydro schemes have an important role to play in the economic development of remote rural areas, especially mountainous ones. Micro-hydro schemes can provide power for industrial, agricultural and domestic uses through direct mechanical power or by the coupling of the turbine to a generator to produce electricity.

The Tunga–Kabiri micro-hydropower scheme, Kenya, showing the power house and penstock

Source: Practical Action / Zul Mukhida

Table 7.1 Classification of hydropower by size

Size	Power output
Large	More than 100 MW and usually feeding into a large electricity grid
Medium	15–100 MW – usually feeding a grid
Small	1–15 MW – usually feeding into a grid
Mini	Above 100 kW, but below 1 MW; either stand-alone schemes or more often feeding into the grid
Micro	From 5 kW up to 100 kW; usually providing power for a small community or rural industry in remote areas away from the grid
Pico	From a few hundred watts up to 5 kW; for domestic use including small-scale agro-processing activities such as threshing, hulling and milling, for battery-charging stations, and for poultry rearing and incubators.

Note: kW (kilowatt) = 1000 watts; MW (megawatt) = 1 million watts or 1000 kW

Scheme components

Figure 7.1 shows the main components and layout of a run-of-the-river micro-hydro scheme. This type of scheme requires no water storage but instead diverts some of the water from the river, channelling it along the side of a valley before it is 'dropped' into the turbine via a penstock.

In the scheme shown in Figure 7.1 the turbine drives a generator that provides electricity for a workshop. The transmission line can be extended to a local village to supply domestic power for lighting and other uses. There are other configurations which can be used, depending on the topographical and hydrological conditions, but all adopt the same general principle.

Water into watts

To determine the power potential of the water flowing in a river or stream it is necessary to determine both the flow rate of the water and the head through which the water can be made to fall. The *flow rate* is the quantity of water flowing past a point in a given time. Typical flow rate units are litres per second or cubic metres per second. The *head* is the vertical height, in metres, from the turbine up to the point where the water enters the intake pipe or penstock.

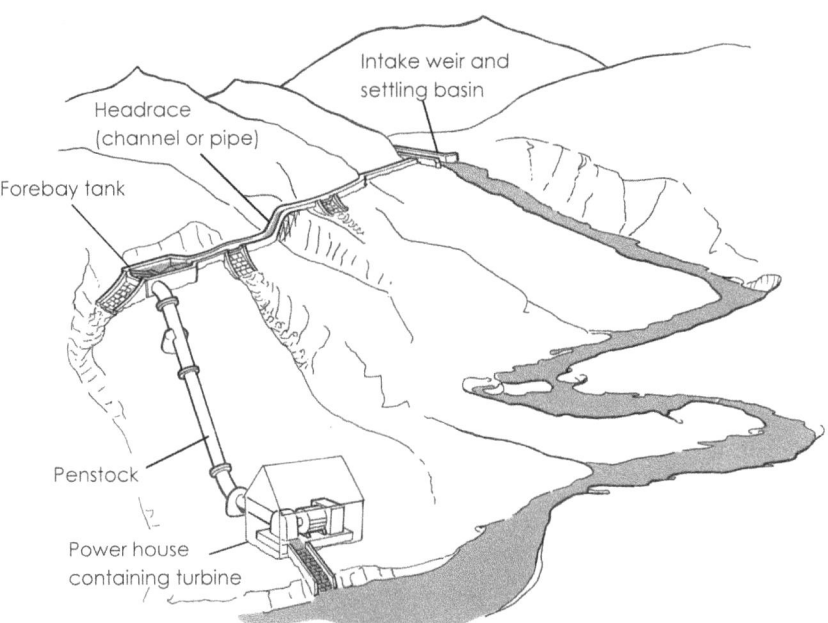

Figure 7.1 Layout of a typical micro-hydro scheme

Source: Practical Action

The potential power can be calculated as follows.

Theoretical power (P) = gravity (g) × flow rate (Q) × head (H)
When g = 9.81 m/s², Q is in m³/s and H in m, then

P = 9.81 × Q × H (kW)

However, energy is always lost when it is converted from one form to another. Small water turbines rarely have efficiencies better than 80 per cent. Power will also be lost in the pipe carrying the water to the turbine, due to friction losses. By careful design, this loss can be reduced to only a small percentage. A rough guide used for small systems of a few kilowatts' rating is to take the overall efficiency as approximately 50 per cent. Thus, the theoretical power must be multiplied by 0.5 (or halved) for a more realistic figure.

Example: A turbine generator set operating at a head of 10 m with flow of 0.3 m³ per second will deliver approximately (9.81 × 0.5 × 0.3 × 10 =) 18 kW of electricity.

It is always preferable to run all equipment at the rated design flow and load conditions, but it is not always practical or possible where river flow fluctuates throughout the year or where daily load patterns vary considerably. If a machine is operated under conditions other than full load or full flow then other significant inefficiencies must be considered. Part flow and part load characteristics of the equipment need to be known for the performance under these conditions to be assessed. Alternatively, an electronic load controller should be used, to ensure optimal loading conditions on the equipment at all times.

Uses of micro hydropower

Depending on the end-use requirements of the generated power, the output from the turbine shaft can be used directly as mechanical power or the turbine can be connected to an electrical generator to produce electricity. For many rural industrial applications shaft power is suitable (for food processing such as milling or oil extraction, sawmill, carpentry workshop, small-scale mining equipment, etc.), but many applications require conversion to electrical power.

For domestic applications electricity is preferred. This can be either:

- provided directly to the home via a small electrical distribution system; or
- supplied by means of batteries which are returned periodically to the power house for recharging – this system is common where the cost of direct electrification is too high due to scattered housing (and hence an expensive distribution system).

Where a generator is used, alternating current (AC) electricity is normally produced. Single-phase power is satisfactory on small installations up to 20 kW, but, beyond this, 3-phase power is used to reduce transmission losses and to be suitable for larger electric motors. An AC power supply must be maintained at a constant 50 or 60 cycles/second for the reliable operation of any electrical equipment using the supply. This frequency is determined by the speed of the turbine which must be very accurately governed.

Suitable conditions for micro hydropower

The best geographical areas for exploiting small-scale hydropower are those where there are steep rivers flowing all year round; for example, the hill areas of countries with high year-round rainfall, or the great mountain ranges and their foothills, like the Andes and the Himalayas. Islands with moist marine climates, such as the Caribbean Islands, the Philippines and Indonesia are also suitable. Low-head turbines have been developed for small-scale exploitation of rivers or irrigation canals where there is a small head but sufficient flow to provide adequate power.

To assess the suitability of a potential site, the hydrology of the site needs to be known and a site survey carried out, to determine actual flow and head data. Hydrological information can be obtained from the meteorology or irrigation department, usually run by the national government. This data gives a good overall picture of annual rain patterns and likely fluctuations in precipitation and, therefore, flow patterns. The site survey gives more detailed information of the site conditions to allow power calculation to be done and design work to begin. Flow data should be gathered over a period of at least one full year where possible, to ascertain the fluctuation in river flow over the various seasons. There are many methods for carrying out flow and head measurements and these can be found in the relevant texts or online resources.

Turbines

A turbine converts the energy in falling water into shaft power. The choice of turbine will depend mainly on the head available and the design flow for the proposed hydropower installation. As shown in Table 7.2, turbines are broadly divided into three groups: high, medium, and low head, and into two categories: impulse and reaction.

Table 7.2 Classification of turbine types

	Head pressure		
Turbine	High	Medium	Low
Impulse	Pelton Turgo Multi-jet pelton	Crossflow Turgo Multi-jet pelton	Crossflow
Reaction	na	Francis Pump as turbine (PAT)	Propeller Kaplan

Source: Harvey, 1993

MICRO HYDROPOWER

The difference between impulse and reaction can be explained simply by stating that the impulse turbines convert the kinetic energy of a jet of water in air into movement by striking turbine buckets or blades – there is no pressure reduction as the water pressure is atmospheric on both sides of the impeller. The blades of a reaction turbine, on the other hand, are totally immersed in the flow of water, and the angular as well as linear momentum of the water is converted into shaft power – the pressure of water leaving the runner is reduced to atmospheric or lower.

Load factor

The load factor is the amount of power used divided by the amount of power that is available if the turbine were to be used continuously. Unlike technologies relying on costly fuel sources, the 'fuel' for hydropower generation is free and therefore the plant becomes more cost-effective if run for a high percentage of the time. If the turbine is only used for domestic lighting in the evenings then the load factor will be very low. If the turbine provides power for rural industry during the day, meets domestic demand during the evening, and maybe pumps water for irrigation in the evening, then the plant factor will be high.

It is very important to ensure a high load factor if the scheme is to be cost-effective and this should be taken into account during the planning stage. Many schemes use a 'dump' load (in conjunction with an electronic load controller – see below), which is effectively a low priority energy demand that can accept surplus energy when an excess is produced e.g. water heating, storage heaters or battery charging.

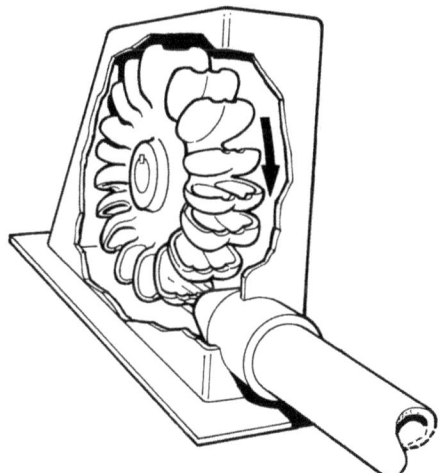

Figure 7.2 An impulse turbine
Source: Practical Action

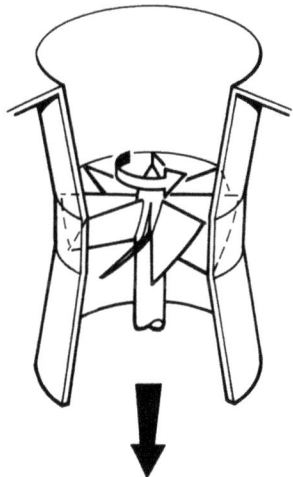

Figure 7.3 A reaction turbine
Source: Practical Action

Load control governors

Water turbines, like petrol or diesel engines, will vary in speed as load is applied or removed. Although not a great problem with machinery that uses direct shaft power, this speed variation will seriously affect both frequency and voltage output from a generator. Traditionally, complex hydraulic or mechanical speed governors altered flow as the load varied, but more recently electronic load controllers (ELC) have been developed which have increased the simplicity and reliability of modern micro-hydro sets. The ELC prevents speed variations by continuously adding or subtracting an artificial load so that, in effect, the turbine is working permanently under full load. A further benefit is that the ELC has no moving parts, and is reliable and virtually maintenance-free. The advent of electronic load control has allowed the introduction of simple and efficient multi-jet turbines, no longer burdened by expensive hydraulic governors.

The economics – cost reduction

Normally, small-scale hydro installations in rural areas of developing countries can offer considerable financial benefits to the communities served, particularly where careful planning identifies income-generating uses for the power.

The major cost of a scheme is for site preparation and the capital cost of equipment. In general, unit cost decreases with a larger plant and with high heads of water. The costs of micro hydropower can be kept low by avoiding some of the more costly aspects of large hydropower. This includes such innovations as:

- using run-of-the-river schemes where possible – this does away with the cost of an expensive dam for water storage;
- locally manufactured equipment where possible and appropriate;
- use of HDPE (plastic) penstocks where appropriate;
- electronic load controller – allows the power plant to be left unattended, thereby reducing labour costs, and introduce useful by-products such as battery charging or water heating as dump loads for surplus power;
- using existing infrastructure – for example, a canal which serves an irrigation scheme;
- siting of power close to village to avoid the need for expensive high-voltage distribution equipment such as transformers;
- using pumps as turbines (PAT) – in some circumstances standard pumps can be used 'in reverse' as turbines, which reduces costs, delivery time, and makes for simple installation and maintenance;
- using motors as generators – as with the PAT idea, motors can be run 'in reverse' and used as generators; pumps are usually purchased with a motor fitted and the whole unit can be used as a turbine/generator set;
- local materials can be used for the civil works;
- use of community labour;

- good planning for a high plant factor (see above) and well balanced load pattern (energy demand fluctuation throughout the day);
- low-cost connections for domestic users (see following section on this topic);
- self-cleaning intake screens – this is fitted to the intake weir to prevent stones and silt from entering the headrace canal, which removes the need for overspill and desilting structures along the headrace canal and also means that, in many cases, the canal can be replaced by a low-pressure conduit buried beneath the ground – this technology is, at present, still in its early stages of dissemination.

Maintenance costs (insurance and water abstraction charges, where they apply) are a comparatively minor component of the total – although they may be an important consideration in marginal economic cases.

Ownership and management

Programmes promoting the use of micro hydropower in developing countries have concentrated on the social as well as the technical and economic aspects of this energy source. Technology transfer and capacity-building programmes have enabled local design and manufacture to be adopted. Local management, ownership and community participation has meant that many schemes are under the control of local people who own, run and maintain them. Operation and maintenance are usually carried out by trained local craftspeople.

Low-cost grid connection

Where the power from a micro-hydro scheme is used to provide domestic electricity, one method of making it an affordable option for low-income groups is to keep the connection costs and subsequent bills to a minimum. Often, rural domestic consumers will require only a small quantity of power to light their houses and run a radio or television. There are a number of solutions that can specifically help low-income households to obtain an electricity connection and help utilities meet their required return on investment. These include:

Load-limited supply. Load limiters work by limiting the current supplied to the consumer to the prescribed value. If the current exceeds that value then the device automatically disconnects the power supply. The consumer is charged a fixed monthly fee irrespective of the total amount of energy consumed. The device is simple and cheap and does away with the need for an expensive meter and subsequent meter reading.

Reduced service connection costs. Limiting load supply can also help reduce costs on cable, as the maximum power drawn is low and so smaller cable sizes can be used. Also, alternative cable poles can sometimes be found to help reduce costs.

Pre-fabricated wiring systems. Wiring looms can be manufactured 'ready to install' which will not only reduce costs but also guarantee safety standards.

Credit schemes. These allow householders to overcome the barrier imposed by the initial costs of grid connection. Once connected, energy savings on other fuels can enable repayments to be made. Using electricity for lighting, for example, incurs a fraction of the cost of using kerosene.

Community involvement. Formation of community committees and co-operatives who are proactive in all stages of the electrification process can help reduce costs as well as provide a better service. For example, community revenue collection can help reduce the cost of collection for the utility and hence the consumer.

Appropriate scale hydropower

In recent years there has been much debate over the appropriate scale of hydropower. Many argue that large hydro is not only environmentally damaging (as large areas of land are flooded) but that there is also a negative social impact where large imported technologies are used. However, small-scale hydropower avoids the negative social and environmental impacts of large hydro. Micro hydro can deliver many social and economic benefits to isolated communities who otherwise would not have access to modern energy.

Further information

Bibliography

Cunningham, P. and Woodfender, I. (2007) 'Microhydro-electric systems simplified', *Home Power,* February/March. Available at: https://homepower.com/articles/microhydro-electric-systems-simplified

Dhanapala, K. and Wijayatunga, P. (n.d.), *Best Practices for Micro-Hydro Development*, Practical Action South Asia, Colombo.

Fraenkel, P., Paish, O., Bokalders, V., Harvey, A. and Brown, A. (1991) *Micro-Hydro Power: A Guide For Development Workers,* Practical Action Publishing, Rugby, IT Power and Stockholm Environment Institute.

Harvey, A. (1993) *Micro-Hydro Design Manual: A Guide to Small-scale Water Power Schemes*, Practical Action Publishing, Rugby.

Khennas, S. and Barnett, A. (2000) *Best Practices for Sustainable Development of Micro Hydro Power in Developing Countries*, DFID, UK, and The World Bank. Available at: www.microhydropower.net/download/bestpractsynthe.pdf

Practical Action (n.d.), *Civil Works Guidelines for Micro Hydropower in Nepal,* Practical Action Nepal and BPC Hydroconsult, Kathmandu.

Rodríguez, L. and Sánchez, T. (2011) *Designing and Building Mini and Micro Hydro Power Schemes: A Practical Guide*, Practical Action Publishing, Rugby.

Segura, J., Rodríguez, L. and Núñez, P. (1995) *Layout and Lining of Canals*, Soluciones Prácticas, Lima.

Smith, N. (1994) *Motors as Generators for Micro-Hydro Power*, Practical Action Publishing, Rugby.

Thake, J. (2000) *The Micro-hydro Pelton Turbine Manual: Design, Manufacture and Installation for Small-scale Hydropower,* Practical Action Publishing, Rugby.

Williams, A. (1995) *Pumps as Turbines – A User's Guide*, Practical Action Publishing, Rugby.

Useful addresses

International Hydropower Association (IHA)
Web: www.hydropower.org
IHA addresses the role of hydropower in meeting the world's growing water and energy needs as a clean, renewable and sustainable technology.

International Network on Small Hydro Power (IN-SHP)
Web: www.inshp.org/main.asp
Since its establishment in 1994, IN-SHP has worked to promote global SHP development with the aim of rural electrification.

International Small-Hydro Atlas
Web: www.small-hydro.com
International Small-Hydro Atlas is an international database that facilitates the development of new small hydro projects of more than 50 kW and less than 10 MW installed capacity. It provides data for potential and developed sites, GIS searching capabilities, country profiles, international contacts for small hydro, and more.

Micro-hydro Centre
Web: www.picohydro.org.uk

Micro hydro website maintained by Wim Klunne
Web: http://microhydropower.net
Contains a useful range of information including case studies, manufacturers, consultants, theory, discussion groups, and downloadable books and manuals

Renewables First
Web: www.renewablesfirst.co.uk
Provides useful information on hydropower design

'Micro hydropower' was written by Alison Doig in 2007 and last updated in 2011. Alison Doig has a PhD in rural energy systems and spent eight years as an energy specialist for ITDG/Practical Action working on micro hydropower, cookstoves and energy policy. Since then she has worked for WWF and as a private consultant, and her current role is as policy adviser at Christian Aid, specializing in climate change, sustainable development and low-carbon energy.

Chapter 8

River turbines

This chapter describes a hydro-kinetic turbine, which uses the flow of a river but runs without a head of water, thus not requiring water to fall any distance. Theoretical power calculations are presented before looking at the practicalities of using small turbines in remote locations at low cost.

Keywords: hydro, renewable, energy, water, hydro-kinetic, micro hydro, river turbines, battery charging

Introduction

Hydro-kinetic turbines are designed to generate electricity solely from the movement of running water in a river, or from tidal currents when used in marine settings, whereas the conventional technique of generating electricity from hydrological energy is done using water from a high position that falls through a head onto a turbine, where water is channelled along canals and pipes in order to make use of its potential energy. This approach is covered in Chapter 7. The use of kinetic energy from river currents is a less common, alternative approach to hydropower where there is zero head. The energy is converted into electricity, or used directly to power a mechanical water pump for irrigation.

Making use of kinetic energy from river currents was a traditional way to mill flour, along with wind power, but was gradually replaced by fossil fuel systems. There is now renewed interest in river current turbines for electricity generation in a wide range of countries. Practical Action has been involved in promoting small-scale turbines to provide electricity to remote villages in the Amazon and to supply water for irrigation in Sudan.

The power available from the river

Most of the principles of this type of turbine are based on wind turbines, as they work in a similar way. The power available (P_a) in watts can be worked out using the following formula.

$$P_a = \tfrac{1}{2} \times C_p \times \rho \times A \times V^3$$

A = area (m²)
ρ = density of water (1000 kg/m³)
V = velocity of water (m/s)
C_p = the power coefficient = 16/27 = 0.592

A river current turbine in Peru

Source: Soluciones Prácticas

The theoretical maximum power available from the river is expressed by the equation above using a power coefficient of 0.592 or 59 per cent efficiency. However, a small-scale river turbine has its own losses which will reduce the power coefficient to around 0.25. The significant aspect to the equation is that the power increases in a cubed relationship to the velocity of the flow of water past the turbine. It is therefore important to find the best flow to get the best power output.

Hydro-kinetic turbines can be classified into two types. The first is the vertical-axis turbine, whose turning axis is perpendicular to stream flow; secondly, the axial turbine, whose rotational axis follows the direction of flow. Vertical-axis turbines are preferable in situations where the flow direction changes, such as in tidal systems. These turbines are designed so that the direction of rotation of the turbine remains the same regardless of the direction of flow.

Putting the theory into practice

For small-scale low-cost river turbines, Practical Action found the second configuration to work due to the simplicity of the design and its greater robustness. It also has the advantage of building on existing knowledge from the small-scale wind turbine projects previously undertaken.

The experience from Peru highlighted the importance of the site of the turbine. Larger rivers are better in providing the right conditions for a turbine but the flow of a river can vary considerably over the year. In a full river the turbine can be located near to the bank and still work in deep water, but during the dry season the water levels can drop and the turbine might hit the bottom. If the bank has a shallow incline then the edge of the water moves away from its originally position as the water level drops and the turbine needs to be moved further towards the middle of the river. It is therefore important to find a site where the water will have a fast flow near to steep banks.

A major hazard for small river turbines in large rivers is debris such as logs or trees that have fallen into the river. These can seriously damage the turbine and incur large costs. Banks can be eroded or become unstable and in extreme cases the river can change its course. These environmental issues can be more challenging than the technical issues.

The design

The small-scale turbine designed by Practical Action focuses on providing battery-charging facilities for remote communities. Typically, a small turbine with a capacity of 200 W could charge four batteries in a day.

Turbine blades

The design of blades is achieved using concepts that apply to wind turbines with the proviso that in this case the machine is subjected to much stronger forces with, of course, a denser medium (water).

In order to calculate the diameter of the rotor the following formula is applied:

$P = \frac{1}{2} \times \rho \times (A) \times V^3 \times C_p \times h$

$A = (\pi \times d^2 / 4)$

$d = \sqrt{(8 \times P / \pi \times \rho \times V^3 \times C_p \times h)}$

d : diameter of turbine rotor (m)
P : power of aero-generator design (W)
ρ : density of water (kg/m³)
V : velocity of river water (m/s)
A : area covered by the turbine (m²)
C_p : power coefficient (no dimensions)
h : efficiency of the generator

$\lambda = U/V_D = \pi \times N \times d/60 \times V_D$
$N = (60 \times \lambda \times V_D / \pi \times d)$

N : velocity of turbine rotor (rpm)
λ : tip speed ratio
U : tangential velocity at the tip of the blade (m/s)
V_D : design velocity (m/s)

The rotor

- three fibreglass blades;
- nominal diameter: 1.75 m;
- turning speed: 45 rpm, at 1 m/s to the speed of river flow;
- two stainless steel supporting plates for the bucket mounting.

The generator

In order to reduce costs, and to be able to rely on locally made technology, Practical Action began by working on the development of a permanent magnet generator. The magnets allowed the speed of generation to be reduced and lowered the cost of the equipment, which itself could be adapted to be a river turbine rotor and, ultimately, tested and built. The generator produces an alternating current which, via a system of rectifying diodes, transforms the voltage to 12 V, and 250 W of power at 360 rpm.

Transmission shaft

A galvanized steel tube 1.5 inches (3.8 cm) in diameter, connected directly to the rotor. This tube is laid inside a second, similar tube, 2.5 inches (6.4 cm) in diameter, which serves as protection and support.

The fan belt and permanent magnet generator

Source: Soluciones Prácticas

Other component parts

Fan belt. An intermediate component connecting the transmission shaft and the generator, it amplifies the speed of rotation.

Control panel. This includes basic measuring instruments and the aforementioned 12 V rectifying diodes.

Floats. The floats can be made in a number of ways based on what is most suitable in terms of the materials available, and could be a boat. In Peru balsa-wood floats made by inhabitants of the village were used, as this was the cheapest option.

Further information

Bibliography

Maldonado Quispe, Francisco (2005) *Diseño de una Turbina de Río para la Generación de Electricidad en el Distrito de Mazán-Región Loreto,* Universidad Nacional Mayor de San Marcos, Lima. Available at: http://sisbib.unmsm.edu.pe/BibVirtualData/monografias/basic/maldonado_qf/maldonado_qf.pdf

Ramírez, Saúl and Escobar, Rafael (2002) *Turbina de Rio: Una Alternativa Energética para la Amazonía* [river turbine: alternative energy in the Amazon]. Hidrored 2/2002, Soluciones Prácticas, Lima.

van Els, R.H., de Oliveira, C., Mantel Dias, A. and Balduino, L.F. (2003) *Turbina Hidrocinética para Poblaciones Aisladas.* [The hydro-kinetic turbine in isolated communities]. Hidrored 1/2003, Soluciones Prácticas, Lima.

Useful addresses

Ampair
PO Box 416, Poole, Dorset BH12 3LZ, UK
Tel: +44 (0)1202 749994, Web: www.ampair.com
Underwater micro-hydro 100 W generator

Thropton Energy Services
Physic Lane, Thropton, Northumberland NE65 7HU, UK
Tel: +44 1669 621288, Web: www.throptonenergy.co.uk

'River turbines' is based on the document produced in Spanish by Giannina Solari of Soluciones Prácticas and translated by Edward Stevens for Practical Action. Giannina Solari is a mechanical engineer working in Soluciones Prácticas (Practical Action Latin America).

Additional information was provided by Teo Sanchez. Teo Sanchez is the Technology and Policy Advisor for Energy at Practical Action. He is a mechanical engineer with substantial experience in small-scale energy delivery working in Peru and the United Kingdom. He studied mechanical engineering at the National University of Engineering, Lima, Peru, and has an MSc from the University of Reading and a PhD from Nottingham Trent University, both in the UK.

Chapter 9
Biogas

This chapter explains what biogas is, how it is produced and the energy that can be obtained. It also looks at the carbon dioxide emissions balance. Technical, social and economic issues are discussed, looking at the different types of digester that are available and what their social impact can be.

Keywords: biogas, renewable, energy, digester, fuel, methane, fermentation, animal waste

Introduction

Biogas is a well-established fuel for cooking and lighting in a number of countries. It is a gas mixture, comprising around 60 per cent methane and 40 per cent carbon dioxide, that is formed when organic materials such as dung or vegetable matter are broken down by microbiological activity in the absence of air, at slightly elevated temperatures (most effective at 30–40°C or 50–60°C). This is the same process as occurs naturally at the bottom of ponds and marshes and gives rise to marsh gas or methane.

China has over 7.5 million household biogas digesters, 750 large- and medium-scale industrial biogas plants, and a network of rural 'biogas service centres' to provide the infrastructure necessary to support dissemination, financing and maintenance. India has about three million household-scale systems installed (Martinot, 2003). Other countries in the South with active programmes include Nepal, Sri Lanka, Kenya, and several countries in Latin America. As carbon emission levels are causing greater concern and as people realize the benefits of developing integrated energy supply options, biogas is becoming an increasingly attractive option.

The biogas process is known as anaerobic (without air) digestion, and provides a clean cooking and lighting fuel that can be produced on a scale varying from a small household system to a large commercial plant of several thousand cubic metres. Biogas can be used for electricity generation and for powering farm equipment. There are two main types of electricity generation equipment.

- *Microturbines* are small gas turbines that burn methane mixed with compressed air. As they burn, the hot pressurized gases are forced out of the combustion chamber and through a turbine wheel, causing it to spin and turn the generator, thus making the electricity.
- *Reciprocating gas engines* have been modified from natural gas engines so that they can handle the larger quantities of carbon dioxide and contaminants that are found in biogas.

The digestion of animal and human waste yields several benefits.

- The production of methane for use as a fuel reduces the amount of woodfuel required and thus reduces desertification.
- The waste is reduced to slurry that has a high nutrient content, making an ideal fertilizer.
- During the digestion process, dangerous bacteria in the dung and other organic matter are killed, which reduces the pathogens dangerous to human health.

Carbon emissions

In some cases, anaerobic digestion is used to produce fertilizer as the main product, and the biogas is merely a by-product which is vented from the digester. This has serious negative environmental impacts as methane is a damaging greenhouse gas. However, when the gas is burnt, it is one of the few energy processes that is 'carbon negative' in that it reduces the amount of greenhouse gas emitted by the raw material (dung emits methane), making it an attractive option for those seeking carbon funding for wide-scale dissemination.

Practical issues

There are several technologies for obtaining biogas.

- The most common is the fermentation of human and/or animal waste, diluted to slurry, in specially designed digesters.
- Where water is scarce, an adapted technology uses a drier mix with high yields and more manageable residues.
- A recent approach using starches from waste foods and grain in much smaller quantities has created a small-scale technology appropriate for both urban and rural communities.
- Where there are no cattle, new technologies show that fuel crops can yield biogas.
- Larger-scale, more recently developed technologies capture methane from municipal waste landfill sites.

When building a biogas digester, certain criteria must be met if it is to be successful.

Technical requirements

- Sufficient raw feedstuffs must be available on a long-term basis and over the whole year, or supplies will be inconsistent and people will lose confidence in the technology.

- The temperature has to be high enough to cause the digestion process to work or additional building work to create a warm environment may make it prohibitively expensive.
- For fixed-dome digesters, the quality of the building materials must be high as the biogas is held under pressure within the dome.
- Skills and know-how are both needed to build and to maintain biogas plants. Many units built in the past have been abandoned for lack of servicing skills.

Social issues

- It is more likely to succeed if there is a market for the fertilizer end-product. This supply chain should be part of the planning stage of biogas introduction.
- Even if the set-up costs are subsidized, those who will use the gas should have some financial stake in the construction or they may not have a sufficient sense of ownership to maintain the plant.
- Handling animal and human wastes is a sensitive cultural issue and even the use of the gas may be unacceptable in some societies.
- Collection of dung may be problematic if the livestock is not held in a fixed place but is allowed to wander freely.
- Promotion and dissemination of the benefits of biogas will be needed if it is to be accepted in the rural areas where feedstock is available.
- The use of human waste appears to be more successful when it is associated with an institution such as a school or a hospital, rather than an individual home.
- NGO involvement can ensure that technologies are appropriate and acceptable to the target community.

Financial/political considerations

- Government promotion and involvement can assist in dissemination. This can be a win–win solution as it provides clean energy and reduces problems associated with waste.
- Private sector investment will support long-term sustainability.
- Set-up costs are relatively high and so may only be affordable to those on higher incomes. Microcredit can be used to reduce this problem. Credit schemes, or well-targeted subsidies, will enable a larger number of people to access biogas technologies and thus stimulate the market. For example, USAID's Nepal Biogas Microfinance Capacity Building Program has established appropriate financial institutions to help continue and sustain the development of the biogas sector in Nepal.

Household-level technologies

The most widespread designs of digester are the Chinese fixed-dome digester and the Indian floating cover biogas digester (shown in Figures 9.1 and 9.2). The digestion process is the same in each digester but the gas collection method is different. In the floating cover type, the water-sealed cover of the digester is capable of rising as gas is produced, where it acts as a storage chamber, whereas the fixed dome type has a lower gas storage capacity and requires good sealing if gas leakage is to be prevented. Both have been designed for use with animal waste or dung.

The waste is fed into the digester via the inlet pipe and undergoes digestion in the digestion chamber. The temperature of the process is quite critical. Methane-producing bacteria operate most efficiently at temperatures of 30–40°C or 50–60°C, and in colder climates heat may have to be added to the chamber to encourage the bacteria to carry out their function. The product is a combination of methane and carbon dioxide, typically in the ratio of 6:4. Digestion time ranges from a couple of weeks to a couple of months, depending on the feedstock and the digestion temperature. The residual slurry is removed at the outlet and can be used as a fertilizer.

From a household perspective, the gas should always be available, so those digesters which allow continuous addition of feedstock which displaces spent feedstock is likely to be the most appropriate and acceptable. Batch systems, which require the physical removal of slurry every few days and the addition of new feedstock, are labour-intensive and disruptive to supply.

Biogas digesters where water is a constraint

This digester, developed by the Central Institute of Agricultural Engineering, Bhopal, India, is a modification of the fixed-dome type and it allows fresh undiluted cattle dung to be used. The modified design requires very little or no water for mixing with the cattle dung, generates about 50 per cent more biogas for each kilogram of dung loaded into the system, and does not require slurry drying time before it can be used as fertilizer.

The main changes to a conventional fixed dome digester are an increase in the bore of the inlet feed, greater reinforcement of the chamber to withstand the higher gas pressures, an enlarged slurry chamber outlet and a smooth widened outlet channel to streamline the flow of the slurry (Shyam, 2001).

Compact biogas digester using waste foodstuffs

For those without cattle or within urban centres, a conventional digester may not be appropriate. The Indian Appropriate Rural Technology Institute (ARTI) has introduced a small biogas digester that uses starchy or sugary wastes as feedstock, including waste flour, vegetable residues, waste food, fruit peelings, rotten fruit, oilcake, rhizomes of banana, canna (a plant similar to a lily but rich in starch), and non-edible seeds. The compact plant is made from cut-down high-density polythene (HDPE) water tanks, which are adapted using a heat gun and standard HDPE piping. The standard plant uses two tanks, with volumes of typically 0.75 m^3 and 1 m^3. The smaller tank is the gas holder and is inverted over the larger one which holds the mixture of decomposing feedstock and water (slurry).

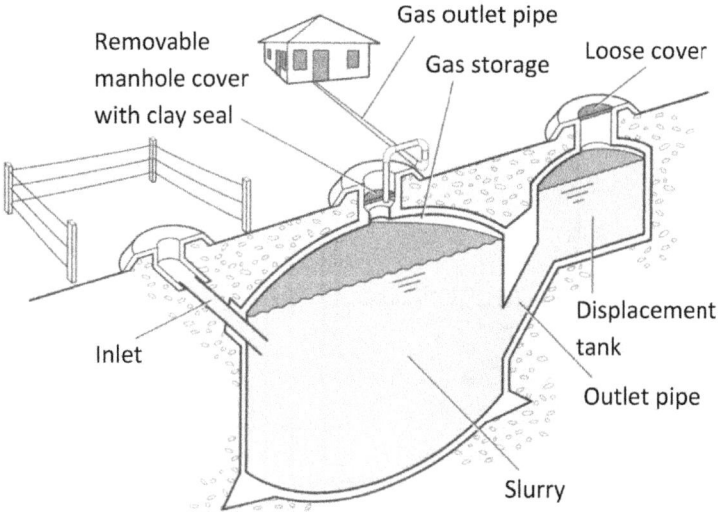

Figure 9.1 Fixed dome digester
Source: Practical Action

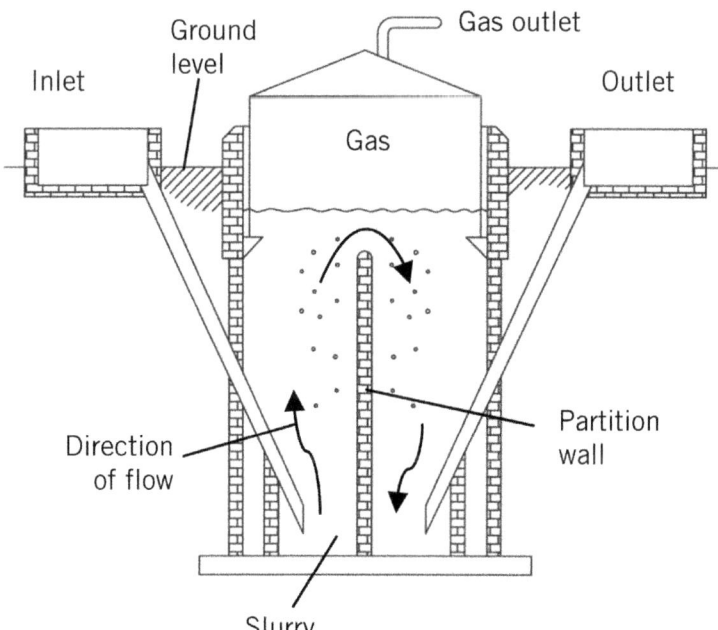

Figure 9.2 Floating cover digester
Source: Practical Action

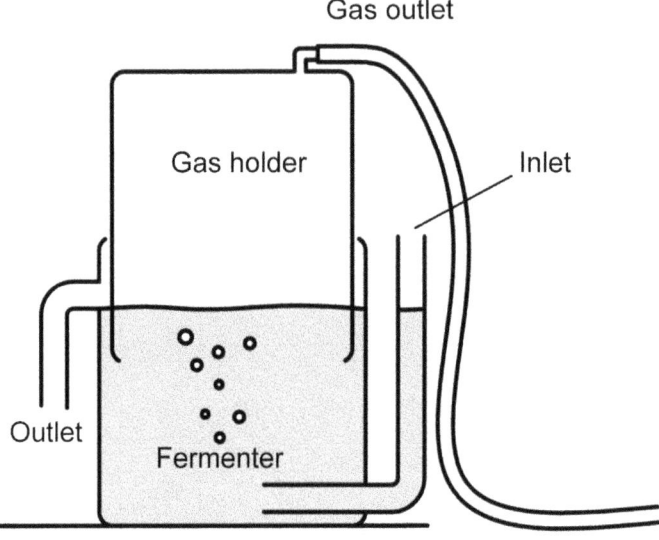

Figure 9.3 Compact biogas digester
Source: Neil Noble / Practical Action

The feedstock must be blended so that it is smooth using a blender powered by electricity or by hand. Two kilograms of such feedstock produces approximately 500 g of methane, and the reaction is completed within 24 hours. An inlet is provided for adding feedstock, and an overflow for removing the digested residue. The digester is set up in a sunny place close to the kitchen, and a pipe takes the biogas to the kitchen (ARTI, 2006).

Larger-scale biogas plants

Industrialized countries commonly use large-scale biogas digesters where animal dung, and increasingly fuel crops, are used as feedstock. Brazil and the Philippines lead the world in crop-based digesters using sugar-cane residues as feedstock. Interest in biogas has been growing in most of the European countries. After a period of stagnation caused by technical and economic difficulties, the environmental benefits of biogas and increasing price of fossil fuel have improved the competitiveness of biogas as a fuel. This has been seen in both small- and large-scale plants in Denmark, Germany (with over 3000 plants producing 500 MW electricity and 1000 MW of heat) and Switzerland, and as a transport fuel in Sweden (where vehicles using biomass were voted environmental cars of the year in 2005). There have been interesting biogas projects in the UK, Ireland, and the Netherlands. Despite this, the use of biogas in Europe is modest in relation to the raw-material potential, and biogas produces only a very small share of the total energy supply.

Several countries are experimenting with dedicated biogas energy crops, such as newly bred grass varieties (Sudan grass and tropical grass hybrids) or biogas 'super maize' developed in France. The crops are developed in such a way that they ferment easily and yield enough gas when used as a single substrate. Biogas crops can be used whole, which allows for the use of far more biomass per hectare.

When produced on a large scale, biogas can be fed into the natural gas grid and enters the energy mix without consumers being aware of the change. A select number of European firms have already begun doing so, while in Germany farmers who generate excess biogas on their farms are using it to generate electricity; in 2005, biogas units produced 2.9 billion kilowatt-hours of electricity (AGEE-Stat, 2010).

India is planning to deal with one of its major problems – air pollution from transport – through the use of compressed biogas (CBG). Since over 70 per cent of the world's long-term (2030) growth in demand for automotive fuels will come from rapidly developing countries like India this is highly relevant and is currently in the research phase (Biopact, 2006).

Uses of biogas

Biogas can be used directly for cooking and lighting, or for heat generation, and for electricity production and fuel for cars. Studies in China have shown that when it is used to heat and light greenhouses it photosynthesizes by increasing the carbon dioxide concentration, so increasing yields. Experiments in Shanxi Province have shown that increasing carbon dioxide four-fold between 6 am and 8 am boosts yields by nearly 70 per cent. A biogas lamp gives both light and warmth to silkworm eggs, increasing their rate of hatching as well as cocooning compared with that obtained with the usual coal heating.

At industrial level, the methane and carbon dioxide mix in biogas can be used to inhibit picked fruit from ripening too early as it inhibits metabolism, thereby reducing the formation of ethylene in fruits and grains. It also kills harmful insects, mould, and bacteria that cause diseases (Kangmin and Ho, 2011).

Table 9.1 shows some approximate equivalents of various energy sources compared to 1 m³ of biogas.

Table 9.1 Biogas fuel equivalents[1]

Energy source	Equivalent to 1 m^3 of biogas
Petrol	0.53–0.75 litre
Diesel	0.48–0.68 litre
Firewood	1.50 kg
Electricity	1.20 kWh
LPG	0.46 kg

1 at 15°C and atmospheric pressure

Source: Biogas Technology Center, Chiang Mai University, and FAO, 1992

Social impacts of using biogas

- Biogas is a clean fuel, thus reducing the levels of indoor air pollution, a major cause of ill-health for those living in poverty.
- Lighting is a major social asset, and already there are estimated to be over 10 million households with lighting from biogas (Martinot, 2003). Improved lighting is associated with longer periods for work or study.
- Where biogas is substituted for woodfuel, there are two benefits: a reduction in the pressures on the forest, and time-saving for those who have to collect wood – usually women and children.
- If a biogas plant is linked to latrines in a sanitation programme, it is a positive way of reducing pathogens and converting the waste into safe fertilizer.
- Where biogas is linked with sales of the resultant fertilizer, it is an excellent source of additional income.
- Fertilizer can be used on crops to increase their yield.
- In China and India biogas plants are produced in great numbers by local artisans. In Kenya, where biogas technology is still in its early stages of dissemination, local manufacturers have been quick to realize the potential and get involved with the production of biogas plants.
- Biogas can be used to generate electricity, bringing with it the possibilities of improved communications: telephone, computer, radio and television for remote communities.
- Fuel produced locally is not as vulnerable to disruption as, for example, grid electricity or imported bottled gas.

Further information

Technical Briefs and other guidelines

Biodigester Construction Manual, AIDG,
   http://practicalaction.org/biodigester-construction-manual
Biogas Digest – Volume 1 Biogas Basics, GTZ,
   http://practicalaction.org/biogas-digest
Using a Biogas Digester, http://practicalaction.org/using-a-biogas-digester
Using Biogas Technology to Solve Pit Latrine Waste,
   http://practicalaction.org/using-biogas-technology-to-solve-pit-latrine-waste-disposal

Bibliography

AGEE-Stat (2010) *Renewable Energy Sources 2010*, Federal Ministry for the Environment, Nature Conservation and Nuclear Safety (BMU), Berlin. Available at: www.bmu.de/files/english/pdf/application/pdf/ee_in_zahlen_2010_en_bf.pdf

Anderson, T., Doig, A., Rees, D., and Khennas, S. (1999) *Rural Energy Services: A Handbook For Sustainable Energy Development*, Practical Action Publishing, Rugby.
ARTI (n.d.) *Biogas Plant: A Compact Digester for Producing Biogas from Food Waste*, www.arti-india.org/content/view/45/52 [accessed December 2011].
Biopact (2006) 'India's bright green idea: compressed biogas for cars', October 25, http://news.mongabay.com/bioenergy/2006/10/indias-bright-green-idea-compressed.html [accessed February 2011].
Chiang Mai University (1992) 'The Biogas Technology Center', Chiang Mai, Thailand, http://grove.ufl.edu/~bests/pdfs/050606%20Chuckree%20Senthong.pdf [accessed 3 August 2012].
Clancy, J. and Rebedy, L. (2000) *Electricity in Households and Microenterprises*, Practical Action Publishing, Rugby.
Fulford, D. (1988) *Running a Biogas Programme: A Handbook*, Practical Action Publishing, Rugby.
Gitonga, S. (1997) *Biogas Promotion in Kenya*, Intermediate Technology Group Kenya, Nairobi.
Gunnerson, C.G. and Stuckey, D.C. (1986) *Anaerobic Digestion: Principles and Practices for Biogas Systems*, World Bank Technical Paper No. 49, Washington, DC.
Harris, P. (n.d.) 'Beginner's guide to biogas', University of Adelaide [website] www.adelaide.edu.au/biogas [accessed December 2011].
Johansen, T.B., Kelly, H., Reddy, A.K.N., and Williams, R. (1993) *Renewable Energy Sources for Fuels and Electricity*, Island Press, Washington, DC.
Kangmin, L. and Ho, M-W. (2011) 'Biogas China', Institute of Science in Society. Available at: www.i-sis.org.uk/BiogasChina.php [accessed December 2011].
Karekezi, S. and Ranja, T. (1997) *Renewable Energy Technologies in Africa*, AFREPEN, Nairobi.
Marchaim, U. (1992) *Biogas Processes for Sustainable Development*, FAO, Rome.
Quaak, P., Knoef, H. and Stassen, H.E. (1999) *Energy from Biomass: A Review of Combustion and Gasification Technologies*, World Bank Technical Paper No. 422, Energy Series, Washington, DC.
Ravindranath, N.H. and Hall, D.O. (1995) *Biomass, Energy and the Environment: A Developing Country Perspective from India*, Oxford University Press, Oxford.
Shyam, M. (2001) 'A biogas plant for the digestion of fresh undiluted cattle dung', *Boiling Point* 47, Autumn, Practical Action, Rugby.
Taleghani, G. and Akbar Shabani Kia, A.S. (2005) 'Technical–economical analysis of the Saveh biogas power plant', *Renewable Energy* 30(3): 441–6.
Van Buren, A. (1979) *A Chinese Biogas Manual*, Practical Action Publishing, Rugby.

'Biogas' was produced by Dr Liz Bates for Practical Action in March 2007 and last updated in May 2012. Liz Bates is a household energy specialist with a specific interest in the alleviation of indoor air pollution through reduction in kitchen stove emissions. Her work includes writing technical briefs, editing documents, and support to household energy projects.

Chapter 10

Liquid biofuels and sustainable development

Issues relating to liquid biofuels, such as ethanol and biodiesel, are discussed, outlining the potential as well as the dangers of their increased use. The basics of processing are covered and some of the uses, such as for stoves, lighting and generating electricity.

Keywords: biofuels, biodiesel, ethanol, renewable, energy, fuel, fermentation, biofuel applications

Introduction

Liquid fossil fuels, such as paraffin (kerosene) and fuel oil, have been with us for many years. Over the past decade, similar fuels made by processing plants, trees and organic waste products have become much more widely available. The rapid growth in the use of biofuels stems from the soaring price of fossil oil, growing concern over security of supply and the environmental impact of fossil fuels. The three main types of liquid fuel looked at here are:

- ethanol, made by fermenting sugar cane, grain, straw, grass and wood;
- biodiesel, made from new or recycled vegetable oils and animal fats (e.g. from palm oil);
- oils made by compressing seeds (such as jatropha oil).

Grown sustainably, biofuels have the potential to alleviate global warming and other negative environmental issues such as the disposal of vast quantities of organic wastes. Used responsibly, biofuels can have a major impact on levels of pollutants, both within the homes of those living in poverty, and in the crowded cities of the developing world.

When biofuels first became widely available they were heralded as the new sustainable way to provide the world with energy. The market in biofuels has grown substantially since 2000, especially the production of ethanol (shown in Figure 10.1), and this trend is expected to continue. More recently, the use of land for growing crops which are solely for energy has led to major issues such as land degradation and famine and, as a result, environmentalists are calling for stricter global controls on production.

Biofuels are produced on an industrial scale almost exclusively by developed countries; the United States and Brazil account for nearly 90 per cent of all ethanol production, with all but 2.8 per cent of the total produced in these two countries plus the EU and China (see Figure 10.2). Biodiesel has dominance in the EU, where most of the fuel is produced by Germany and France. However, the United States, Brazil and Argentina have also rapidly increased production of biodiesel in the last couple of years (see Figure 10.3).

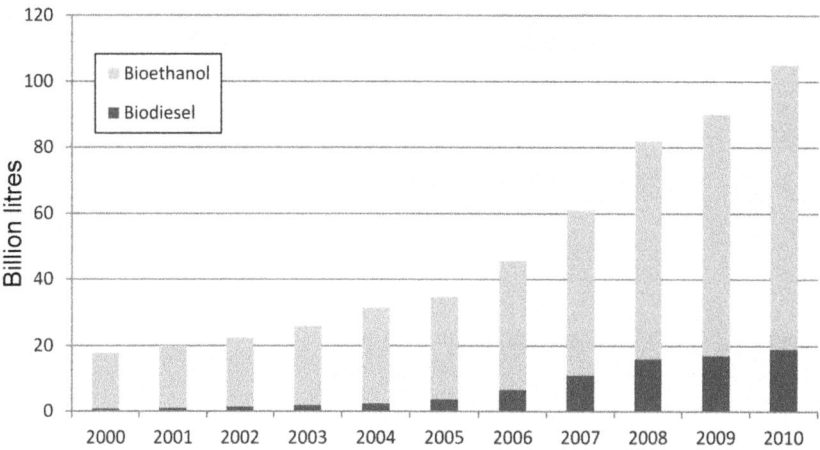

Figure 10.1 Global biofuel production
Source: REN21, 2011

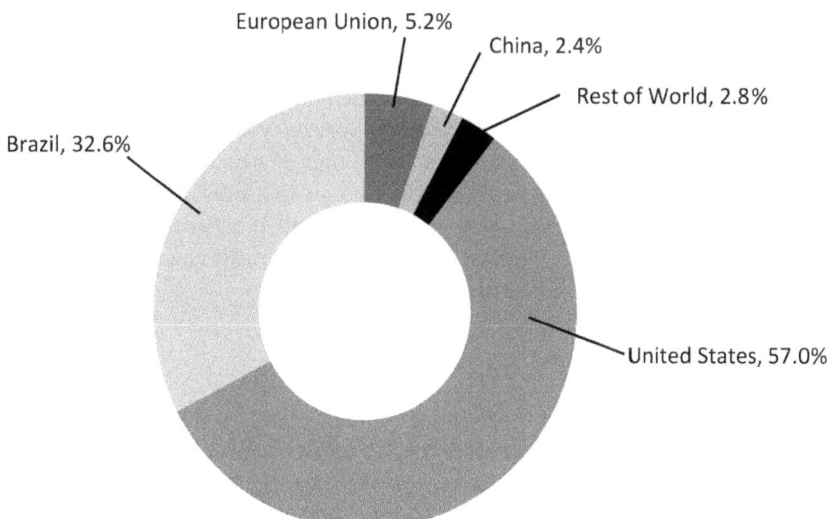

Figure 10.2 2010 ethanol production by region
Source: REN21, 2011

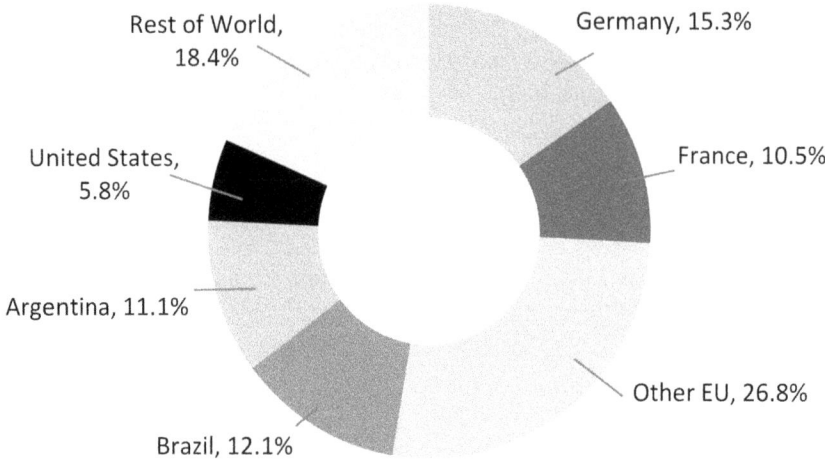

Figure 10.3 2010 biodiesel production by region
Source: REN21, 2011

Growing sugar-based biofuels

Sustainable cultivation

Of all opportunities for renewable energy from plantation biomass, sugar cane makes the most sense. In many African countries under colonial economies, sugar was produced (mainly for export to Europe and beyond) while the residues, such as molasses, were dumped into the rivers, leaching the oxygen from the water and destroying life. These sugar factories were very dirty operations and many still are. Developing a market for ethanol can form part of a beneficial reuse strategy that can clean sugar factories and use the residues that previously had no value and were dumped. This adds value to local farming practices. With advanced technologies now available, biomass such as trees and grasses can also be used as feedstocks for ethanol production.

Large-scale cultivation

The problems associated with growing biofuels stem from the industrialized world trying to grow its way out of its dependence on fossil fuels, and the huge profits that are to be made from growing palm oil and other high-energy crops. In the United States, which is seeking to reduce its dependence on overseas oil products, subsidies may mask the costs of spraying both pesticides and fertilizers, and using large-scale farm machinery.

Growing and harvesting oil-bearing seeds

A vast variety of oil plants originate in the tropics and subtropics. Many oil-bearing plants, whose oils are often toxic to humans, grow on low-grade land or in marginal locations unsuitable for food crops. Some of these plants are cultivated on waste land to prevent further erosion and to inhibit desertification. Use of these oils for energy provision does not need to compete with food production. Examples of these oil plants are the physic nut tree (*Jatropha curcas L.*), the castor oil plant varieties (*Ricinus communis L.*) and the babassú palm (*Orbignya phalerata Mart.*), among others.

Some oil plants are grown with food plants as shade trees or to provide barriers against animals; jatropha is not eaten by animals so it can be used in this way. Recently, jatropha has been grown as an oil-seed crop, but the resultant yields have proved disappointing (Franken, 2010).

Biofuels and the environment

Biofuels can both benefit and destroy the environment, depending on the ways in which they replace fossil fuels. When fossil fuels burn, they add to the levels of greenhouse gases in the environment. Where these are replaced by renewably grown biofuels, the carbon dioxide is re-absorbed by the plants as they grow, and energy does not need to be expended in transporting fossil fuels over long distances. Thus, using clean and sustainable practices, biofuels can benefit the environment, and add economic value to local communities.

However, where large plantations have been planted in previously rainforested areas, this benefit is completely overwhelmed by the damage done to the environment by burning the forest, which is an essential global 'sink' for these greenhouse gases. Indiscriminate culling of forests has led, for instance, to oil-palm plantations covering 5.3 million hectares of Indonesia (2004). There is 'well researched evidence that careless development of oil palm is destroying forests, drying out peat-swamps, wiping out endangered species, polluting air and waterways, driving climate change, dispossessing indigenous peoples and immiserating the rural poor' (Colchester and Chao, 2011). This large-scale, non-regulated approach can lead to land degradation as the soil is leached of its nutrients. After 25 years, lands are often so leached of nutrients that they are abandoned and become scrubland where few plants will grow.

The main use of biofuels is in the automotive sector and again there are both benefits and disadvantages. Ethanol can reduce carbon emissions through improved combustion and reduced reliance on fossil fuels, but access to an alternative source of fuel may increase supply, thereby keeping the price of fuel down, encouraging more people to use private transport and so resulting in higher overall emissions.

Within the household context, however, biofuels have a positive impact on the lives of those living in poverty. The kitchen environment remains the place where most women in the developing world spend most of their time. Alleviating kitchen smoke improves health, saves money, improves women's status, and saves very large amounts of time. In areas prone to drought, it can also help save the external environment by reducing pressures on trees and other vegetation used for fuel.

Processing biofuels from raw feedstock

Bioethanol

As shown in Figure 10.4, the raw feedstock, such as sugar-cane residue, is ground up to a small size and the active ingredients in the feedstock react with dilute sulphuric acid, breaking down into a mixture of simple sugars, cellulose enzymes which are grown at this stage, and cellulose. An alternative approach is to add enzymes rather than make them.

The cellulose is further reacted with the cellulose enzymes to form glucose. Both the simple sugars and the glucose are fermented with yeasts or bacteria, forming ethanol, with carbon dioxide as a by-product. The non-reactive parts of the sugar cane remain as a woody waste which can be used as a fuel to drive the process.

Micro distilleries, such as the pilot micro distillery at Viçosa Federal University, Brazil, are well-suited to making ethanol in local districts. These advanced technologies can produce ethanol of the same quality as huge industrial plants, and are less likely to be involved in the international transport fuel sector. They are therefore suitable for manufacturing ethanol for stoves, close to the point of use. This type of micro distillery can be shared by groups of farmers, allowing them

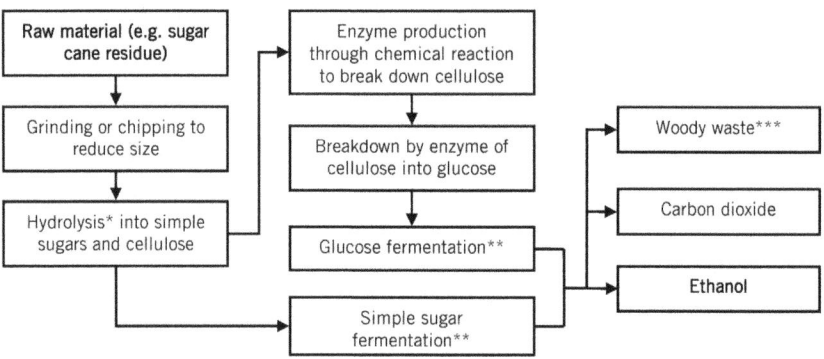

Figure 10.4 Bioethanol production process

Note:
* chemical reaction caused by addition of dilute sulphuric acid on the active ingredients in the feedstock
** series of chemical reactions that convert sugars through action by yeast or bacteria
*** woody waste can be used to power the process

to add value to their sugar crops. Micro-distillery technology is currently (2012) in use in Brazil, and is being transferred to Ethiopia for a Project Gaia stoves programme there.

Biodiesel

For biodiesel, the raw oil is reacted with an excess of alcohol (often methanol or ethanol) in the presence of a catalyst (often potassium or sodium hydroxide). The products of this reaction are crude biodiesel and crude glycerine, with the excess alcohol which is recovered and reused.

Plant oils for energy

Plant oils are currently under scrutiny as safe and renewable household fuels. They do not burn under normal room conditions, and designs are being tested that burn cleanly. However, none to date has reached a fully commercial stage.

Plant oils from seeds and nuts

The methods of extraction are similar to those used for food processing. Practical Action has extensive information on this, including a technical brief specifically on oil extraction.

Local oil production strengthens decentralized supply, provides employment and income opportunities for the local population, and promotes sustainability. The oilcake, a by-product of the oil processing, can be used either as fodder or as high-quality fertilizer. In general, all plant oils which are liquid at ambient temperatures can be used as cooking fuel (Stumpf, 2002).

Uses of biofuels

The main use of the fuel alcohols (ethanol and methanol) globally is within the transport sector, with bioethanol replacing petrol (gasoline) and biodiesel replacing diesel. Governments have been keen to promote the use of such fuel for transport purposes as they can draw increased revenue from taxing its sale as motor fuel. In the longer term, unregulated planting of energy crops for the

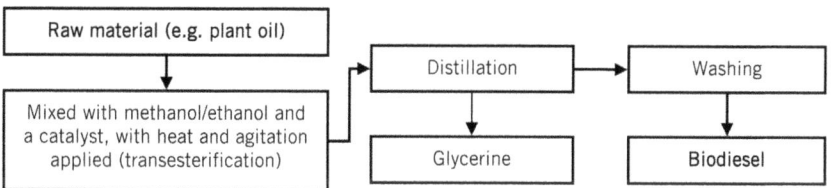

Figure 10.5 Biodiesel production process

transport sector is likely to lead to land degradation, social unrest and famine. As well as for food, vegetable oils (particularly those that are inedible, such as jatropha) are used in products such as soap and cosmetics.

For cooking and lighting within the home, biofuels are hugely beneficial, as they can be burnt much more cleanly. With over 4000 people dying each day globally as a result of indoor air pollution from cooking smoke, provision of a clean local alternative fuel in those countries that are most affected can have a substantial effect on the alleviation of poverty. Clean cooking promotes health, safety and quality-of-life benefits, especially for women and children. Developing countries with ideal climates for the rapid growth of fuel crops such as sugar cane are often the same ones suffering the greatest burden of disease and death from indoor air pollution.

Biofuel stoves have been shown to be effective in reducing or eliminating the practice of gathering biomass, which can be especially critical where deforestation and desertification are making it difficult to collect, or in conflict situations where women fuel-gatherers are particularly vulnerable to assault.

Technologies for ethanol

Liquified petroleum gas (LPG) is a fossil-fuel derivative of two large energy industries: natural gas processing and crude oil refining. In 2010 global LPG production reached a record high of 249 million tonnes, with the Asia-Pacific region growing almost 4.6 per cent. The global LPG market is expected to increase substantially in the next decades as LPG markets in Asia and Africa grow (World LP Gas Association).

Any downturn in the supply of these two fossil fuels will lead to a reduction in the availability of supplies and consequent price increases. Other clean fuels are needed for cooking both to complement LPG in countries which are not oil-rich, and to supply clean energy to those currently using biomass in traditional cooking stoves and on three-stone fires.

Stoves

Ethanol is easily and safely handled, leading to its widespread use for recreational markets such as yachting and other outdoor activities. Recently, robust, low-cost stoves have been designed specifically for households in the developing world. A good example is the CleanCook Stove, a stainless steel stove which burns cleanly and has safety features designed particularly for the household:

- A fuel tank holds the ethanol in a special absorptive fibre so that it cannot spill out.
- Fuel is denatured so that it cannot be ingested.
- The tank is not pressurized, so there is no risk of explosion.
- The burner flame is easily adjusted or extinguished by means of a simple regulator.

This stove is finding a market in both Africa and Latin America (Project Gaia). Another stove, which runs on lower-grade ethanol and produces a lower heat output, is the Nari stove, designed by the Nimbkar Agricultural Research Institute (NARI). Fuel made from ethanol, ethanol gel, was for a time considered to be a useful way of providing ethanol. However, more recent studies have shown that the gelling agent makes the stove burn at a lower temperature, and it burns less cleanly. Ethanol gel is now used in the industrialized world more for aesthetic purposes, and as a back-up fuel in some countries, such as South Africa.

Lighting

Ethanol can be used to produce a clean, bright light. Petromax, who produce the Britelyt Methanol/Ethanol/Alcohol Lantern, makes the best-known of these lanterns. This lantern is manufactured in the United States (Britelyt). The Nimbkar Agricultural Research Institute (NARI) in the Philippines has developed the Noorie lantern which burns very cleanly, producing light equivalent to that from a 100 W incandescent light bulb. If the top of the lamp is removed, an attachment can allow the lantern to be used for small amounts of cooking.

Ethanol and biodiesel for transport

Ethanol is a high octane fuel and has replaced lead as an octane enhancer in petrol. When ethanol is blended with petrol, oxygen is added to the fuel mixture so it burns more completely and reduces polluting emissions. Ethanol fuel blends are widely sold in the United States. The most common blend is 10 per cent ethanol and 90 per cent petrol (E10). Vehicle engines require no modifications to run on E10. However, only flexible-fuel vehicles can run on up to 85 per cent ethanol and 15 per cent petrol blends (E85).

Brazil and the United States are the two major ethanol producers. In Brazil, a government decision was taken more than twenty years ago to make the country self-sufficient in energy. By 2005, the number of cars sold that could run on both ethanol and petrol exceeded those sold that could use petrol alone. As of 2010, ethanol accounted for 41.5 per cent of all light-duty transport fuel consumption in Brazil (REN21, 2011).

The global production of biodiesel is currently on the same scale as ethanol, although several countries in the EU have developing infrastructures and increasing production figures. The EU currently accounts for over 50 per cent of biodiesel production, the majority of which is used in blends with petro-diesel through national blending quotas. Although biodiesel production figures have slowed in both the United States and the EU since the worldwide economic crisis of 2008, it is estimated that by 2020 biodiesel could represent as much as 20 per cent of all on-road diesel used in Brazil, Europe, China and India (Biodiesel 2020, 2008). Additionally, many new biodiesel plants are required to fuel a growing industry in Argentina. Another key future development will be 'advanced biodiesel' derived from 'second-generation' feedstocks such as micro-algae; whilst these fuels are not currently commercially viable, significant research is being conducted into their scaling up, and they have the potential to supply a significant proportion of all biodiesel by 2050 (IEA, 2010).

Technologies for plant oil

Stoves

A wide variety of plant oils have potential for cooking, including coconut, jatropha, soybean, corn, peanut, cotton, sunflower and many more. There are currently a number of designs at prototype stage. However, to date, none has reached the commercial stage. Those piloting the use of a plant-oil stove say that it cooks faster than a traditional kerosene stove and the cost of fuel tends to be lower.

Electricity generation

There are several types of vegetable oils that can be used to generate electricity in adapted diesel engines. For example, in the Amazon region, around 1000 small power plants (<500 kW) use diesel to supply electricity to small towns and villages, whilst huge plantations of oil palm are cultivated for biofuels for export (Coelho, 2005).

Diesel is not subsidized, and the electricity is often too expensive for households to buy. Diesel engines adapted to burn vegetable oil are being tested in demonstration units in isolated villages. Where they have been installed, they have been found to be socially beneficial – particularly where the communities are very isolated. Such projects are potential candidates for carbon finance as they replace diesel, which can lead to lower costs for both installation and fuel.

In Mali, according to the news agency Reuters, some 700 communities have installed biodiesel generators powered by oil from the *Jatropha curcas* plant. The Malian government is promoting cultivation of the inedible oilseed bush to provide electricity for lighting homes, running water pumps and grain mills, and other critical uses. Mali hopes to eventually power all of the country's 12,000 villages with affordable, renewable energy sources. Jatropha has the additional benefit that it stabilizes soil in areas prone to erosion, and is used as a medicinal plant.

Mali is seeking to boost the standard of living of its 80 per cent rural population and to reduce migration from impoverished rural areas through electricity for light, air conditioning, vaccine storage and media (Herro, 2007).

In rural Cambodia a scheme was piloted whereby jatropha nuts, which grow wild along the roadside, are collected by local villagers and traded with the local shopkeeper in exchange for goods to the value of the nuts. The nuts are bought by the oil-producer for this traded price and small oil-expellers located in some of the villages will be used to make jatropha oil. The oil has been supplied to businesses that charge batteries and run standard diesel generators. The oilcake from the crushed seeds is used as cattle feed.

Future for biofuels

Biofuels are here to stay. Used responsibly, they can reduce the levels of pollutants affecting the planet, and can provide clean air in homes fit for human habitation. As new technologies become mainstreamed (such as fuel cells for the auto industry) it is to be hoped that these new technologies can improve the quality of life for those communities currently living in poverty, and provide a route to clean development for future generations.

Further information

Technical Briefs and other fact sheets

Oil Extraction, http://practicalaction.org/oil-extraction
Production and Use of Jatropha,
    http://practicalaction.org/production-and-use-of-jatropha-1

Bibliography

Coelho, S.T. (2005) *Biofuels – Advantages and Trade Barriers*, United Nations Conference on Trade and Development, http://unctad.org/en/docs/ditcted20051_en.pdf

Colchester, M. and Chao, S. (2011) *Oil Palm Expansion in South East Asia: Trends and Implications for Local Communities and Indigenous Peoples*, Forest Peoples Programme and SawitWatch, http://pdf.usaid.gov/pdf_docs/PNADS887.pdf

Emerging Markets Online (2008) *Biodiesel 2020: Global Market Survey, Feedstock Trends and Forecasts*, 2nd edn, Emerging Markets Online, Houston, TX. Summary available at: www.emerging-markets.com/PDF/Biodiesel2020Study.pdf [accessed 7 August 2012].

Franken, J. (2010) *Jatropha: Retrospective and Future Development*, International Conference on Jatropha Curcas, November 1–2, FACT Rijksuniversiteit Groningen, Netherlands.

Guarte, R.C. (2006) 'LPG alternative seen in new plant oil stove', posted 3 May, http://neda8.evis.net.ph/eddnews/stove/stove.htm

Herro, A. (2007) *Eye on Mali: Jatropha Oil Lights Up Villages*, WorldChanging – Change Your Thinking, www.worldchanging.com/archives/006814.html

IEA (2010) *Sustainable Production of Second-Generation Biofuels: Potential and Perspectives in Major Economies and Developing Countries*, International Energy Agency.

Mongabay.com (2007) 'Massive oil palm expansion planned by Indonesia's richest man', May 8, http://news.mongabay.com/2007/0508-palm_oil.html

O'Brien, C. (2006) 'Introducing alcohol stoves to refugee communities', *Boiling Point* 52, Practical Action. Available at: www.hedon.info/docs/BP52-7-OBrien.pdf [accessed 4 September 2012].

REN21 (2011) *Renewables 2011: Global Status Report*, Renewable Energy Policy Network for the 21st Century, Paris.

Stumpf, E. and Mühlbauer, W. (2002) 'Plant-oil cooking stove for developing countries', *Boiling Point* 48, Practical Action. Available at: https://practicalaction.org/docs/energy/docs48/bp48_pp37-38.pdf

World LP Gas Association (n.d.) 'About LP gas', www.worldlpgas.com/about-lp-gas [accessed 4 September 2012].

Useful addresses

Approtech Oil Processing Technologies, www.approtec.org/tech_oil.shtml
BluWave Ethanol Stove,
    www.hedon.info/docs/BluwaveEthanolStoveAssessment.pdf
Britelyt lanterns, www.britelyt.com/lanterns.htm
Energypedia Biofuels, https://energypedia.info/index.php/Portal:Biofuel
HEDON Household Energy Network, www.hedon.info/goto.php/Ethanol
Nimbkar Agricultural Research Institute (NARI), Low-concentration ethanol stove
    for rural areas in India: www.nariphaltan.org/ethstove.pdf
Project Gaia, www.projectgaia.com/
Renewable Energy Policy Network for the 21st Century, www.ren21.net/
Renewable Fuels Association (United States), www.ethanolrfa.org
Symposia, www.fact-foundation.com

'Liquid biofuels and sustainable development' was produced by Dr Liz Bates for Practical Action in March 2007, and last updated by Dr Liz Bates and Martin Bounds in May 2012. Liz Bates is a household energy specialist. Her work includes writing technical briefs, editing documents, and support to household energy projects. She has a particular interest in ethanol, and has been involved in projects using this fuel.

Martin Bounds was an EWB intern at Practical Action. He studied mechanical engineering at the University of Bath where he had a particular interest in the production of biofuels in relation to engine performance.

Chapter 11
Biomass as a solid fuel

This chapter covers bioenergy sources such as wood, crop residue and dung. It looks at the basic combustion theory and then at some of the technical aspects of fuel processing, such as charcoal production and briquetting as well as stove design. Related issues of commercialization, environment and social impact, especially for women, are highlighted.

Keywords: biomass, renewable, energy, solid fuels, crop residue, woodfuel, charcoal production

Introduction

Biomass is the term used to describe all the organic matter produced by photosynthesis that exists on the earth's surface. The source of all energy in biomass is the sun, the biomass acting as a kind of energy store. To make use of biomass for our own energy needs we can simply burn it in an open fire to provide heat for cooking, warming water or warming the air in our home. More sophisticated technologies have been developed for extracting this energy and converting it into useful power and heat in more efficient and convenient ways.

Until relatively recently it was the only form of energy that was used by humans and it is still the main source to fulfil the domestic energy needs of more than half the world's population.

The extraction of energy from biomass is split into three distinct categories:

- *Solid biomass* such as trees, crop residues, animal waste (although not strictly a solid biomass source, it is often included in this category), household or industrial residues can be combusted to provide heat. Often the solid biomass will undergo physical processing such as cutting, chipping, briquetting, etc. but retains its solid form.
- *Biogas* is obtained by anaerobically (i.e. in an air-free environment) digesting organic material to produce a combustible gas known as methane. Animal waste and municipal waste are two common feedstocks for anaerobic digestion. See Chapter 9 for more details.
- *Liquid biofuels* are obtained by subjecting organic materials to one of various chemical or physical processes to produce a combustible liquid fuel. Biofuels such as vegetable oils or ethanol are often processed from industrial or commercial residues such as bagasse (sugar-cane waste remaining after the sugar is extracted) or from energy crops grown specifically for this purpose.

Biofuels are often used in place of petroleum-derived liquid fuels (see Chapter 10, 'Liquid biofuels and sustainable development', Noble, 2012).

Here we look at the use of solid biomass fuels and their associated technologies.

Biomass use

Solid biomass is widely used in developing countries, mainly for cooking, heating water and domestic space heating. Biomass is available in varying quantities throughout the developing world – from densely forested areas in the temperate and tropical regions of the world, to sparsely vegetated arid regions where collecting wood for use as household fuel is a time-consuming and arduous task.

In past decades the threat of global deforestation provided a focus for the efficient use of biomass (as well as introducing alternative fuels) in areas where woodfuel was in particular shortage. Although domestic fuelwood users can suffer greatly from the effects of deforestation, it often arises because of land clearing for agricultural use or for commercial timber.

There have been many programmes aimed at developing and disseminating improved stove technologies to reduce the burden, primarily borne by women, of fuelwood collection as well as reducing health risks associated with smoke from burning fuelwood. Technologies have also been introduced to help with the processing of biomass to improve efficiency, allow for easy transportation, or to make it more usable.

Crop and industrial biomass residues are now widely used in many countries to provide centralized medium- and large-scale production of process heat for electricity or other commercial end-uses. There are several examples in Indonesia of timber-processing plants using wood-waste-fired boilers to provide heat and electricity for their own needs, and occasionally for sale to other consumers. There are also small-scale options for using crop residues.

Combustion theory

For solid biomass to be converted into useful heat energy it has to undergo combustion. Although there are many different combustion technologies available, the principle of biomass combustion is essentially the same for each. There are three main stages to the combustion process.

Drying. All biomass contains moisture, and this moisture has to be driven off before combustion proper can take place. The heat for drying is supplied by radiation from flames and from the stored heat in the body of the stove or furnace.

Pyrolysis. The dry biomass is heated and when the temperature reaches between 200°C and 350°C the volatile gases are released. These gases mix with oxygen and burn producing a yellow flame. This process is self-sustaining as the heat from the burning gases is used to dry the fresh fuel and release further volatile gases. Oxygen has to be provided to sustain this part of the combustion process. When all the volatiles have been burnt off, charcoal remains.

Oxidation. At about 800°C the charcoal oxidizes or burns. Again oxygen is required, both at the fire bed for the oxidation of the carbon and, secondly, above the fire bed where it mixes with carbon monoxide to form carbon dioxide which is given off to the atmosphere.

It is worth bearing in mind that all the above stages can occur within a fire at the same time, although at low temperatures only the first stage will be under way and later, when all the volatiles have been burned off and no fresh fuel has been added, only the final stage will be taking place.

Combustion efficiency varies depending on many factors: fuel, moisture content and calorific value of fuel, etc. The design of the stove or combustion system also affects overall thermal efficiency and Table 11.1 gives an indication of the efficiencies of some typical systems (including non-biomass systems for comparison).

Table 11.1 Efficiencies of some biomass energy conversion systems

Type of combustion technology	Percentage efficiency
Three-stone fire	10–15
Improved wood-burning stove	20–25
Charcoal stove with ceramic liner	30–35
Sophisticated charcoal-burning stove	up to 40
Kerosene pressure stove	53
LPG gas stove	57
Steam engine	10–20

Source: adapted from Kristoferson, 1991

Improved stoves

Much of the research and development work carried out on biomass technologies for rural areas of developing countries has been based on the improvement of traditional stoves. This was initially in response to the threat of deforestation but has also been focused on the needs of women to reduce fuel collection times and improve the kitchen environment by smoke removal. There have been many approaches to stove improvement, some carried out locally and others as part of a wider programmes run by international organizations. Figure 11.1 shows examples of successful improved stove types.

Some of the features of these improved stoves include:

- a chimney to remove smoke from the kitchen;
- an enclosed fire to retain the heat;
- careful design of pot holder to maximize the heat transfer from fire to pot;
- baffles to create turbulence and hence improve heat transfer;
- dampers to control and optimize the air flow;
- a ceramic insert to minimize the rate of heat loss;

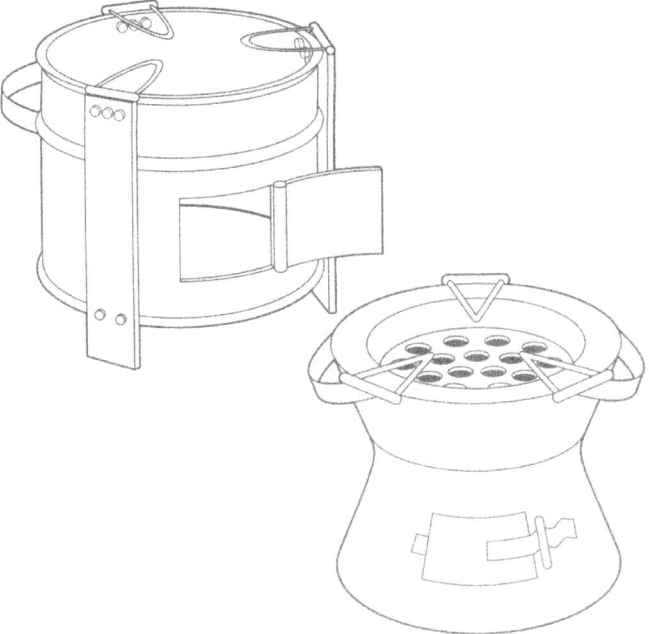

Figure 11.1 A traditional metal stove and the improved Kenya Ceramic Jiko
Source: Practical Action

- a grate to allow for a variety of fuel to be used and ash to be removed;
- metal casing to give strength and durability;
- multi-pot systems to maximize heat use and allow several pots to be heated simultaneously.

Improving a stove design is a complex procedure which needs a broad understanding of many issues. Involvement of users in the design process is essential to gain a thorough understanding of the users' needs and requirements for the stove. The stove is not merely an appliance for heating food (as it has become in Western society), but it often acts as a social focus, a means of lighting and space heating. Tar from the fire can help to protect a thatched roof, and the smoke can keep out insects and other pests. Cooking habits need to be considered, as well as the lifestyle of the users. Light charcoal stoves used for cooking meat and vegetables are of little use to people who have staple foods such as ugali (cornmeal commonly made from maize flour), which require large pots and vigorous stirring. Fuel type can differ greatly; in some countries cow dung is used as a common fuel source, particularly where wood is scarce. Cost is also a major factor among low-income groups. Failing to identify these key socio-economic issues will ensure that a stove programme will fail. The function of an improved stove is not merely to save fuel.

> **Box 11.1 Local manufacture of improved stoves**
>
> Since 1982 the Kenya Ceramic Jiko (KCJ), an improved charcoal-burning stove aimed at the urban market, has been developed and manufactured by large numbers of small producers. The KCJ has two main components: metal and fired clay. Both these parts are made by entrepreneurs, the metal part (cladding) being made by small-scale enterprises or individual artisans, while the clay part (liner) is manufactured by slightly larger and more organized enterprises or women's groups. The KCJ is sold by the artisans directly to their customers or through commercial outlets such as retail shops and supermarkets. The stove was initially promoted heavily to develop the market, by the NGO Kengo and by the Kenyan Ministry of Energy, through the mass media, market demonstrations and trade fairs. As a result of this substantial promotion, there were soon more than 200 artisans and micro-enterprises manufacturing some 13,600 improved stoves every month. It is estimated that there are 700,000 such stoves in use in Kenyan households. This represents a penetration of 16.8 per cent of all households in Kenya, and 56 per cent of all urban households in the country.
>
> *Source:* Walubengo, 1995

Charcoal production

Charcoal production is the most common method for processing wood to make it cleaner and easier to use as well as easier to transport, but charcoal does not increase the total energy content of the fuel – in fact the energy content is decreased. Charcoal is often produced in rural areas and transported for use in urban areas.

The process can be described by considering the combustion process discussed above. The wood is heated in the absence of sufficient oxygen which means that full combustion does not occur. This allows pyrolysis to take place, driving off the volatile gases and leaving charcoal (carbon). The removal of the moisture means that the charcoal has a much higher specific energy content than wood. Other biomass residues, such as millet stems or corn cobs, can also be converted to charcoal.

Charcoal is produced in a kiln or pit. The fuel to be carbonized is stacked in a pile and covered with a layer of leaves and earth. Once the combustion process is under way the kiln is sealed, and then only once the process is complete and cooling has taken place can the charcoal be removed. A simple improvement to the traditional kiln is also shown in Figure 11.2. A chimney and air ducts have been introduced which allow for a sophisticated gas and heat circulation system and with very little capital investment a significant increase in yield is achieved. See also Chapter 12, 'Charcoal production' (Noble, 2012).

The Brazilian NGO Pro-Natura has developed a process based on the continuous carbonization of renewable biomass, savannah weeds, reeds, straw of wheat or rice, cotton and corn stems, rice or coffee husk, and bamboo to produce green charcoal.

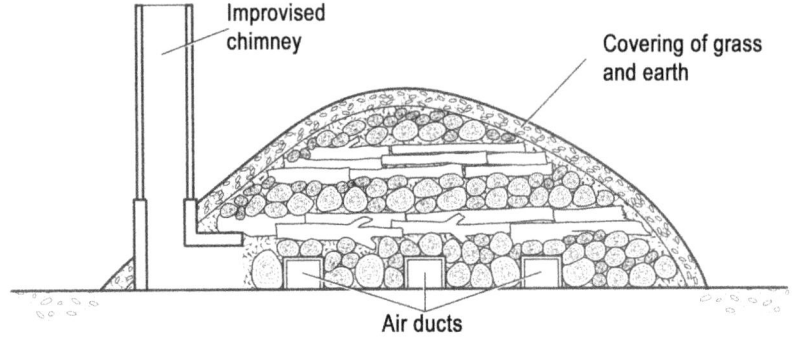

Figure 11.2 Improved charcoal kiln found in Brazil, Sudan and Malawi

Briquetting

Briquetting is carried out on many materials to make them more suitable to be used as an energy source. Nearly all biomass has the potential to be briquetted into a hard stable fuel that has a high energy density and will provide more consistent combustion and improved storage and transportation.

The important factors in making briquettes are the ash content or non-combustible components and the moisture content. The raw materials that are commonly made into briquettes and pellets include:

- wood and sawdust;
- crop waste such as rice husk, cotton stalks;
- bagasse from sugar cane.

Although briquetting is often a large-scale commercial activity, most waste biomass can be used as a fuel source either by directly briquetting or through the production of charcoal that is then briquetted on a small scale. Binders used for direct briquetting include starch paste, cellulose from woody material, cow dung, and clay, which can be extruded or formed into balls by hand.

One example of briquetting sawdust with a binding agent in Malaysia first carbonizes the sawdust, then uses starch as a binder. The starch paste is made in a separate cooking tank.

Charcoal = 73%
Starch = 5%
Calcium carbonate = 2%
Water = 20%.

These charcoal briquettes can be made with a low-pressure mould.

Research by Chardust Ltd into making charcoal briquettes from various crop wastes including sisal waste concluded that carbonizing sisal was technically quite difficult in respect to regulating the temperature, resulting in non-homogeneous carbonization but once the sisal waste had been carbonized it was relatively easy to produce briquettes. These were made by producing a paste of carbon dust and water, and then combined with 15 per cent clay. Briquettes made without a binder are partially carbonized or not carbonized at all. The drawback is that the pressure has to be increased and the equipment used is more complicated.

Most organic waste materials can be burnt directly without being briquetted beforehand. Sawdust stoves and rice husk stoves are relatively common; see the Practical Action Technical Brief, Stoves for Rice Husk and Other Fine Residues.

Dung collection

Many poor families in rural and urban areas use animal dung as a fuel source or collect dung as their source of income. In Bangladesh, people traditionally collect dung, make it into cakes, and sell them to commercial markets. The traditional collectors of dung are teenage girls from poor families. They bring back dung to their homes and convert it into round cakes and cone-like sticks before drying it in the open air.

Box 11.2 Dung as a traditional fuel

Dung is considered to be one of the best fuels for the traditional mud stove for the following reasons:

- It burns slowly.
- It cooks fast.
- It generates powerful heat compared to other sources of fuel found locally.
- It is easy to store
- It produces less toxicity.

Problems related to dung as a fuel are:

- There is a scarcity of dung.
- Cattle owners do not permit collection from their fields.
- As dung is being dried there is a risk that it could be stolen.
- It burns faster than wood when it is not properly compressed.

The alternative approach to using cow dung and other animal waste is the biogas digester which will produce gas that can be used for cooking and lighting or to generate electricity. This will cost more money to implement and maintain.

Commercial utilization of biomass

There are several technologies which employ direct combustion of unprocessed or semi-processed biomass to produce process heat for a variety of commercial activities. The most common is the simple furnace and boiler system which raises steam for such applications as electricity generation or beer brewing. Biomass is also used to provide direct heat for brick burning, lime burning and cement kilns. Rice husk has been used as a heating source for drying in Sri Lanka. See the Technical Brief, Anagi Tray Dryer. The advantage of using biomass is that it can be sourced locally, thereby avoiding shortages associated with poor fuel supply networks and fluctuating costs.

Biomass energy and the environment

Concern for the environment was one of the major inspirations for early research and development work on improved stoves. One of the greatest paradoxes of this work is that, the more that is learnt about people, fuel and cooking, the more it is realized how little was understood about the environment and the implications concerning domestic energy use. Initially, one environmental concern dominated the improved stoves work – saving trees. Today, this issue is considerably downplayed as time has brought a clearer understanding of the true causes of deforestation. At the same time, other environmental issues have become dominant.

Climate change due to the emission of greenhouse gases, especially carbon dioxide with relation to stoves and combustion of biomass, is a complex issue. Wood is carbon-neutral over long time periods but not on a smaller timescale. Therefore, fuel-efficient stoves can reduce CO_2 emissions. Large-scale combustion of biomass is only environmentally feasible if carried out on a sustainable basis. For obvious reasons continual large-scale exploitation of biomass resources without care for its replacement and regeneration will cause environmental damage and also jeopardize the fuel source itself.

Household smoke and the related health issues are now being given greater consideration. These micro environmental needs are often as complex as the broader environmental concerns and this is reflected in the fact that no single improved stove design can meet the needs of a wide and diverse range of peoples.

Women, woodfuel, work and welfare

For resource-poor women the working day stretches from dawn to long after dark. The pressures on women's time are heavy; cooking and fuel collection are among the most arduous of their tasks. The effects of inhaling biomass smoke during cooking are receiving attention from researchers; chronic bronchitis, heart disease, acute respiratory diseases and eye infections have been linked with smoky interiors, but the impacts of fuel shortage on cooking and nutrition are scarcely noticed.

As fuel shortages make extra demands on time and energy, women are driven to various coping strategies. More time spent collecting fuel can mean less time growing or preparing food so that quality and quantity of food diminish. Malnourished women become more vulnerable to smoke pollution which damages their lungs, eyes, children and unborn babies. But improved stoves can cook faster and burn fuel more efficiently, which lowers levels of exposure to biomass smoke and releases time for other activities. Adapting kitchen design can also help remove smoke from the cooking area.

Greater technology choice can help to emancipate women from drudgery and give them more control over precious resources. In some places cooking is a particularly time-consuming task, so an improved stove which cooks faster may be a source of delight. Elsewhere, fuel management strategies by women save more fuel than carefully planned stove programmes. Stove technologists can offer choices, but decisions about household energy technologies should be left with those who do the cooking.

Women design and manufacture improved cookstoves

Source: Practical Action / Simon Ekless

Further information

Technical Briefs and other guidelines

Anagi Tray Dryer, Practical Action South Asia,
   http://practicalaction.org/anagi-tray-dryer
Kenyan Biomass Waste Conversion Potential, Hedon,
   www.hedon.info/docs/Kenyan_Biomass_Waste_Conversion_Potential.pdf
Producer Gas for Power Generation,
   http://practicalaction.org/producer-gas-for-power-generation
Stoves for Rice Husk and Other Fine Residues,
   http://practicalaction.org/stoves-for-rice-husk-and-other-fine-residues,
Using a Biogas Digester, http://practicalaction.org/using-a-biogas-digester

Bibliography

Abbott, Vivienne, Heyting, Clare and Akinyi, Rose (1995) *How to Make an Upesi Stove: Guidelines for Small Business*, Practical Action Kenya, Nairobi. http://practicalaction.org/how-to-make-an-upesi-stove

Ashley, Caroline and Young, Peter (1994) *Stoves for Sale: Practical Hints for Commercial Dissemination of Improved Stoves*, ITDG, FAO, IDEA, GTZ, FWD.

Gitenga, Steven (1997) *Appropriate Mud Stoves in East Africa*, Practical Action Kenya, Nairobi. http://practicalaction.org/mud-stoves-in-east-africa

Kristoferson, L. and Bokalders, V. (1991) *Renewable Energy Technologies: Their Applications in Developing Countries*, Practical Action, Rugby.

Muchiri, Lydia and Sengendo, May (1999) *Appropriate Household Energy Technology Development*, Practical Action Kenya, Nairobi. http://practicalaction.org/appropriate-household-energy-technology-development

Noble, Neil, ed. (2012) *A Handbook of Small-scale Energy Technologies: Practical Answers*, Practical Action Publishing, Rugby.

Stewart, B et al. (1987) *Improved Wood, Waste and Charcoal Burning Stoves,* Practical Action Publishing, Rugby.

Theuri, D., Owala, H., Okello, V., Hada, J., Bikram Mall, M., Kanta Parajuli, T., Shrestha, R., Hood, A., Osman, A., Saleem, M., Bates, L., Bruce, N. and Pope, D. (2005) *Smoke, Health and Household Energy Volume 1: Participatory Methods for Design, Installation, Monitoring and Assessment of Smoke Alleviation Technologies,* Practical Action Publishing, Rugby.

Theuri, D., Owala, H., Okello, V., Hada, J., Bikram Mall, M., Kanta Parajuli, T., Shrestha, R., Hood, A., Osman, A., Saleem, M., Bates, L., Bruce, N. and Pope, D. (2006) *Smoke, Health and Household Energy Volume 2: Researching Pathways to Scaling Up Sustainable and Effective Kitchen Smoke Alleviation*, Practical Action Publishing, Rugby.

Walubengo, Dominic (2005) 'Commercialization of improved stoves: the case of the Kenya Ceramic Jiko', in B. Westhoff and D. Germann (eds), *Stove Images*, pp. 48–9, Sozietät für Entwicklungsplanung (SfE), Frankfurt. Available at: www.giz.de/Themen/de/dokumente/en-stove-images1-1995.pdf

Warwick, H. and Doig, A. (2003) *Smoke – the Killer in the Kitchen,* Practical Action Publishing, Rugby.

Useful addresses

ARTI Appropriate Rural Technology Institute
2nd Floor, Maninee Apartments, Opposite Pure Foods Co., Dhayarigaon, Pune 411 041, India
Web: www.arti-india.org/content/view/42/52/
Briquetted charcoal from sugar-cane waste

Boiling Point
Web: www.hedon.info/BoilingPoint
A practitioner's journal for those working with household energy and stoves

Chardust Ltd
PO Box 24377, Nairobi, Kenya
Web: www.chardust.com
Low-cost charcoal briquettes

Food and Agriculture Organization (FAO)
Via delle Terme di Caracalla, 00100 Rome, Italy
Web: www.fao.org

The HEDON Household Energy Network
Web: www.hedon.info
An informal forum dedicated to improving social, economic, and environmental conditions in the South, through the promotion of local, national, regional and international initiatives in the household energy sector.

The Legacy Foundation
4886 Hwy. 66, Ashland, OR 97520, USA
Web: www.legacyfound.org

Pro-Natura International
15, avenue de Ségur, 75007 Paris, France
Tel: +33 (0)1 53 59 97 98 , Web: www.pronatura.org
Starting up in Brazil in 1985, by 1992 Pro-Natura had become one of the first 'Southern' NGOs to be internationalized following the Rio Conference.

'Biomass as a solid fuel' was written by Practical Action and last updated in May 2012.

Chapter 12
Charcoal production

This chapter introduces the basics of carbonization and kiln efficiency, before looking at kiln technology with some practical examples from Sudan and Kenya. Issues of production and use, including briquetting, are covered.

Keywords: charcoal, renewable, energy, biomass, fuel, charcoal production, carbonization, briquetting, kiln

Introduction

Charcoal is widely used as a domestic fuel for cooking in many towns and cities in developing countries as it burns more cleanly and is easier to use than wood. Small-scale charcoal production is labour-intensive. It can be divided into five different stages of operation:

- growing the fuel;
- harvesting the wood;
- drying and preparing the wood for carbonization;
- carbonizing the wood to charcoal;
- screening, storage and transport to warehouse or distribution point.

However, there can be a detrimental effect on the surrounding environment if demand for fuel increases beyond what can be supplied on a sustainable basis.

Here we look at some of the approaches in the production of charcoal on a small scale in developing countries where efficiencies can be greatly improved through the adoption of better techniques and equipment.

Although wood is the most common fuel source many other sources have been tried, including agricultural waste such as millet stems and corn cobs as well as coconut shell. These biomass materials are made up of cellulose, lignin, and volatile substances and water. During the production process the volatile components are driven off and the cellulose and lignin are decomposed. The process is divided into the following stages.

- *Combustion*: oxygen supply is high and temperature rises from ambient to over 500°C, and the fire is established; the oxygen supply is reduced after the firing hole is closed and the temperature drops to about 120°C.
- *Dehydration*: free water is driven out at a reduced temperature of about 100°C and the kiln gives out thick, white steam.

- *Exothermic reaction:* when the wood has dried, temperatures rise to about 280°C and the wood begins to break down into charcoal, water vapour and other chemicals; the smoke at this stage is yellow, hot and oily and the temperature is maintained by controlling the air flow through holes and vents to help burn more wood.
- *Cooling:* when carbonization is complete, the kiln cools to below 100°C and charcoal can be removed for further cooling.

The process of carbonization is greatly dependent on the kiln temperature, the moisture content of the wood used (the drier the better), the skill of the producer and the condition of the wood (lignin content). The success of the carbonization process is the efficiency of a kiln, defined as the mass of charcoal obtained as a percentage of the mass of wood initially put into the kiln:

$$E_k = M_c/M_w$$

where

E_k = kiln efficiency
M_c = mass of charcoal produced
M_w = mass of wood put into the kiln

Strictly speaking, this is the recovery efficiency, whereas the conversion efficiency includes the charcoal fines (rejects) that may not be packaged for sale due to their small size. Both efficiencies are calculated on wet/dry air or oven dry basis. For example, if a piece of wood weighing 100 kg has 20 kg of free water, then the actual weight of the wood is 80 kg. The moisture content is thus:

$$\text{Moisture content (MC)} = \frac{\text{Mass of water}}{\text{Mass of wood (dry or wet)}} \times 100\%$$

Wet or dry air basis: MC = 20/100 × 100%
= 20%

Oven dry basis: MC = 20/80 × 100%
= 25%

Thus if a kiln produces, say, 10 kg of charcoal, then the kiln conversion efficiency (E_{kc}) at:

Wet or dry air basis: E_{kc} = 10/100 × 100%
= 10%

Oven dry basis: E_{kc} = 10/80 × 100%
= 12.5%

Now, assuming 5 per cent of the charcoal ends up as fines or dust that cannot be packaged, 0.5 kg of charcoal will remain in the kiln and the kiln recovery efficiency, E_{kr}, can be calculated as follows:

Wet or dry air basis: E_{kr} = (10–0.5)/100 × 100%
= 9.5/100 × 100%
= 9.5%

Oven dry basis: E_{kr} = (10–0.5)/80 × 100%
= 9.5/80 × 100%
= 11.9%

Normally, kiln efficiencies are based on the simplest conditions, that is, conversion efficiencies on dry air basis, as these are the easiest to measure and calculate. Most small-scale charcoal production relies on partial combustion of the wood charge to provide the heat necessary for carbonization hence yields depend heavily on the moisture content (Stassen, 2002).

Kilns

Traditional charcoal production is an acquired skill. The most critical factor in the efficient conversion of wood to charcoal is the careful operation of the kiln. Wood must be dried and carefully stacked to allow an even flow of air through the kiln and sufficient time for reactions to take place. If kilns are not operated correctly, yields can be half the optimum level.

Traditional kilns

Much charcoal for domestic consumption in developing countries is produced in pit kilns (holes dug in the ground), or in mound kilns (piles of wood stacked on the ground and covered with soil), by farmers and landless labourers. Yields (weight of charcoal/weight of wood) from pits vary from less than 10 per cent to over 25 per cent. An improved traditional kiln is shown in Chapter 11 'Biomass as a solid fuel' (Noble, 2012).

Brick and concrete kilns

Kilns made of bricks can be more efficient than earth mounds, can be operated all year round and have longer lifetimes than metal or mud kilns, and are less susceptible to poor operator practices. However, the high-grade charcoal that they produce may not be acceptable to domestic users, since it is difficult to ignite. Switching to large, efficient kilns has many economic and social implications, as most charcoal is still produced by farmers and landless peasants who, under normal circumstances, might not be able to benefit from the switch and, indeed, might suffer from it.

Brick kilns are ideal for replacing traditional kilns when consistent high-quality charcoal is required in large quantities. The throughput of a battery of seven beehive kilns, for example, is around 15,000 m^3/year. However, the construction of such kilns requires a relatively high level of brick-building skills, as well as a supply of bricks. This restricts the scope of such kilns in many countries, but in areas where they can be cheaply built and maintained, they have proved to be a very effective method of charcoal-making. The beehive kilns cost approximately $200–300 (£128–192 at a conversion rate of $1.00 = £0.64) with yields of up to 35 per cent of input wood.

One of the major advantages of the brick kiln over earth kilns of similar size is that their carbonization cycle is much quicker. Typically, a 50 m^3 brick kiln has a carbonization cycle of 8–10 days, whereas that of the comparable earth kiln is at least twice as long. Moreover, the labour involved in operating the brick kiln is very much less than that required to construct and manage the earth kiln. Furthermore, the operation of the brick kiln is generally much simpler than that of the earth kiln: workers can be trained in its use relatively easily and shortages of skilled labour are not likely to be a constraint on production.

Brick kilns, however, are usually permanent structures and are therefore only suitable in locations where there is a supply of wood within easy transport distance and sufficiently large to last the working life of five or more years of the kilns.

Portable steel kilns

Portable steel kilns are in the form of a cylinder with a conical top. The kiln breaks down into three components which are designed to be easily rolled along the forest floor to new burn areas or to be transported by truck. Portable steel kilns have a small output: the annual production from a typical demountable kiln with a volume of 7 m^3 is in the range of 100–150 tonnes. They are not, therefore, particularly suitable for areas where there is a need for high-volume production. Their ideal application is where the source of wood is dispersed and charcoal-making is carried out on a relatively small scale.

The advantages of the portable steel kiln are that it requires less labour than the small earth kiln and has a generally greater yield of more consistent and higher quality charcoal. It is also much quicker: the total carbonization cycle with a 7 m^3 demountable steel kiln is 3–4 days; with a similar size earth mound, the cycle is likely to be 10–14 days.

The mobile steel kiln, like the brick kiln, has the substantial advantage over the earth kiln in that training in its use is very easy. The steel kiln can, therefore, be used even in areas where there is no tradition of charcoal-making. The major disadvantage of the portable steel kiln over traditional kilns is its increased cost: even with local manufacture, this is about $1,000 (around £640) and, in many places, considerably more. Given a working life of two to three years, it can be very difficult to justify economically in areas where labour costs and charcoal prices are low.

Using a portable metal kiln to produce charcoal in Sudan

Source: Practical Action Sudan

Box 12.1 Charcoal from mesquite shrubs in Sudan

Practical Action has made use of portable steel kilns in Sudan to make charcoal from mesquite (*Prosopis*) which is a perennial woody plant that can grow in the arid conditions of Sudan. It was promoted in the 1970s and 1980s as a source of fuelwood, pods for fodder and as a way of stabilizing soil in efforts to combat desertification. However, it has caused problems when unmanaged as it spreads into areas of grazing and farming land and has become very difficult to control. The situation has resulted in the plant being declared a noxious weed in Sudan and there is a programme of eradication. Charcoal production using mesquite is part of this eradication programme that can also provide an opportunity for some to improve their income.

The project (part of the Biomass Technology Group at the Energy Research Institute in the 1980s) uses mobile metal kilns based on the design for carbonization of cotton stalk which is similar in design to a Mark V metal kiln (a well-known charcoal kiln) but of smaller volume and weight to allow for easy transportation, with a volume of about 2 m^3. Its nominal carbonization efficiency is around 25 per cent.

The metal kiln can be fabricated locally at a low cost as it can be made from empty oil barrels which are purchased in Kassala market. The metal kiln's additional advantage over the traditionally used earth-mound kiln is that it enables fine charcoal to be made from the small branches of the mesquite shrub.

Mini charcoal kilns

The mini charcoal kiln is used in many locations. One design of mini charcoal kiln is described by Rao (1984): a kiln that is constructed from an oil drum, based on a traditional design from the Philippines used to process coconut shell. It yields over 30 per cent high-grade charcoal from an 80 kg charge of firewood.

One person could operate a batch of about 10 oil-drum kilns, producing up to 250 kg of high-quality charcoal per day. The major drawback of this type of kiln will be its short lifetime, but where there are cheap oil barrels and a good market for high-quality charcoal, it could be a profitable small business.

Charcoal briquettes

Briquettes made of agricultural waste can compete with traditional woodfuel if they are of sufficient quality and are priced correctly. This allows the conversion of low-grade residues to marketable fuels. The work by MIT D-Lab in producing charcoal briquettes is described in the Practical Action Technical Brief, Fuel from the Fields: Charcoal from Agricultural Waste.

Experience in Gambia and elsewhere has shown that residue and charcoal briquettes may not burn well in existing stoves. See *Boiling Point* special issue on Briquettes 1989/90.

Box 12.2 Biomass wastes for charcoal in Kenya

In Kenya a study in 2004 looked into using biomass waste to make charcoal briquettes. The study encompassed criteria such as availability, conversion potential, fuel quality and enterprise potential.

The study considered the total quantity available across the country as the most important parameter for fuel production. It was also important to consider any annual or seasonal variations in supply and any pre-existing or competing uses. Finally, the lower the ash content of the biomass waste, the better the quality of fuel production.

Based on this analysis, the following biomass wastes showed potential commercial viability based on their availability within Kenya:

- bagasse;
- sawdust;
- coconut husk;
- coffee husk;
- wattle bark;
- macadamia nut shell.

The study showed that all of these apart from coconut husk have high potential to form the basis of a viable briquetting business in the country.

Source: HEDON, 2004

Issues of production and use

Charcoal is important in terms of energy and economies within most African countries. The production of charcoal employs a considerable number of people in rural areas. However, charcoal users as a group are most strongly exposed to carbon monoxide (CO), followed by wood users. Charcoal use also results in high volumes of carbon dioxide (CO_2) emissions contributing to global warming.

Increasing end-use efficiency requires the promotion of improved stoves. Traditional stoves are normally made by the informal sector; models with higher heat transfer efficiencies should be developed in collaboration with end-users and stove producers, and manufactured by the private sector.

Inefficiencies inherent in the production and use of charcoal, rapid urbanization, and the preference of urban dwellers for charcoal place a heavy strain on local wood resources.

This led to Practical Action investigating the potential for fuel substitution in a number of countries, including Sudan and Kenya, where households were helped to switch from wood and charcoal use to LPG and kerosene. Financial loans helped people cover the costs of converting as this was seen as the dominant constraint. Introducing LPG or kerosene reduces the particle pollutants, which result in improved long-term health benefits when compared with traditional wood or charcoal use for cooking.

Kerosene and LPG are affordable for many upper- and middle-level households, but further improvements in pricing and delivery (particularly of LPG) are required to enable households lower on the income scale to make the switch away from traditional fuels. See Practical Action Technical Brief, Kerosene and Liquid Petroleum Gas (LPG).

Electricity is not a potential substitute for woodfuel. Although electricity is affordable and practical in many areas for lighting, communications, and possibly for refrigeration, few households, rural or urban, will be able to afford to cook with electricity if it is priced at cost-reflective tariffs.

Conclusion

Improved charcoal kilns require some capital outlay but also require better understanding and control of the carbonization process. Drying of wood, better stacking methods, and better process control, in combination with a chimney to force inverted draught, can greatly increase carbonization efficiency. However, they take more time and effort to prepare the kiln and control the carbonization process.

In areas where wood is freely available, traditional charcoal makers may not have an incentive to improve their production and may use several traditional kilns. Increasing the efficiency of charcoal production requires regulatory measures, systematic training, and demonstration programmes.

Further information

Fact sheets

Charcoal of Simple Kiln Systems, GTZ (now GIZ), Eschborn
Charcoal Production Using a Transportable Metal Kiln, Natural Resources Institute, Chatham Maritime
Construction of a Transportable Charcoal Kiln, Natural Resources Institute, Chatham Maritime
Construction of Charcoal Kilns Built with Locally Manufactured Bricks, Natural Resources Institute, Chatham Maritime
Construction, Installation and Operation of an Improved Pit-Kiln for Charcoal Production, Natural Resources Institute, Chatham Maritime
Fuel from the Fields: Charcoal from Agricultural Waste, https://practicalaction.org/fuel-from-the-fields-charcoal-from-agricultural-waste
Kerosene and Liquid Petroleum Gas (LPG), http://practicalaction.org/kerosene-and-liquid-petroleum-gas-lpg

Bibliography

Boiling Point (1989/1990) Special issue on Briquettes, Practical Action, Rugby.
Boutette, Michael and Karch, Ed (1984) *Charcoal: Small-scale Production and Use,* GTZ (now GIZ), Eschborn.
FAO (1983) *Fuel from the Fields: Charcoal from Agricultural Waste. Simple Technologies for Charcoal Making,* FAO Forestry Paper No. 41, Rome. Available at: www.fao.org/docrep/x5328e/x5328e00.htm
Foley, Gerald (1985) *Charcoal Making in Developing Countries Technical Report No. 5,* Earthscan.
HEDON (2004) *The Use of Biomass Wastes to Fabricate Charcoal Substitutes in Kenya – Feasibility Study,* Chardust Ltd and Spectrum Technical Services, Nairobi. Available at: www.hedon.info/docs/Kenyan_Biomass_Waste_Conversion_Potential.pdf
Noble, Neil, ed. (2012) *A Handbook of Small-scale Energy Technologies: Practical Answers,* Practical Action Publishing, Rugby.
Rosillo-Calle, F. (1996) *The Charcoal Dilemma: Finding a Sustainable Solution for Brazilian Industry,* Practical Action Publishing, Rugby.
Stassen, H.E. (2002) 'Developments in charcoal production technology', FAO, Rome. Available at: www.fao.org/docrep/005/y4450e/y4450e11.htm
Stewart, Bill et al. (1987) *Improved Wood Waste and Charcoal Burning Stoves: A Practitioner's Manual,* Practical Action Publishing, Rugby.

Useful addresses

Deutsche Gesellschaft für Internationale Zusammenarbeit (GIZ) GmbH
PO Box 5180, 65726 Eschborn, Germany
Web: www.giz.de/en

Pro-Natura International
15, avenue de Ségur, 75007 Paris, France
Tel: +33 (0)1 53 59 97 98, Web: www.pronatura.org
The head office of Pro-Natura International is currently in Paris, with country offices in Brazil, USA, Ghana, Nigeria and the UK.

Pro-Natura UK
Tubney House, Abingdon Road, Tubney, Oxon OX13 5QL, UK
Tel: +44 (0)1865 241044
In an attempt to reduce deforestation, Pro-Natura has developed green charcoal. This technological innovation, using agricultural residues and unused biomass, produces an environmentally friendly and economically competitive alternative to wood.

Sustainable Energy
Cambrian Buildings, Mount Stuart Square, Cardiff CF10 5FL, UK
Tel: +44 (0)29 20 408990, Web: www.sustainable-energy.co.uk
A small private energy consultancy and technology development company in the UK. Sustainable Energy has completed the development of a prototype charcoal-making system, an innovative system that makes charcoal and uses the by-product gases to generate electricity. The development of the prototype has been part supported by the DTI.

'Charcoal production' was written by John Ndegwa for Practical Action in 2010 and last updated in June 2012. John Ndegwa was an EWB intern at Practical Action. He studied mechanical engineering at the University of Manchester, UK.

Chapter 13
Energy for rural communities

This chapter looks mainly at electricity delivery for remote communities in developing countries, looking at ways to optimize the implementation and use of the electricity supply by involving rural people in establishing a system and increasing commercial use at times of low demand. It also gives information on the approaches to low-cost electrical installation in homes and mini-grid development.

Keywords: renewable, energy, electricity, end use, electrification, mini grid, low-cost electrical installation

Introduction

Rural areas in developing countries have limited access to all types of services, including health, clean water supplies, communication and roads. This lack of access is also true for energy services. Nearly a third of the world's population do not have access to grid electricity and the majority of these people live in rural areas of developing countries. Most have no hope of being connected to a mains electricity grid in the foreseeable future, despite the political pressure on governments to increase grid connectivity. Given the choice, and the money, most people would opt to switch to electricity.

Household electricity consumption

Electricity consumption shows large variations, depending on climate, culture, reliability of supply, and location. Generally, rural households in developing countries have very low consumption, with the primary uses being lighting, radio and television.

The load factor is a measure of actual energy used compared to the maximum possible energy available for use. For small-scale energy delivery it is important to know the load factor and the peak energy demand in order to extract the best possible use from the energy system. In most cases the load factor within the rural setting is below 0.2, but peak demand can often exceed capacity.

Where lighting is the only significant use of electricity, monthly consumption tends to be in the range of 10 to 20 kWh. Two 40-watt incandescent bulbs used for five hours each night, for example, have a monthly consumption of 12 kWh. A radio-cassette player or a small fan can be used for ten hours each day for an additional consumption of 10–15 kWh per month for each appliance. A small colour TV used for six hours a day will add a further 10 kWh a month. A family could accommodate all these uses easily within a consumption range of 50–60 kWh a month. A refrigerator uses about 50 kWh and a freezer around 100 kWh a month. Ideally, extra demand would occur during off-peak periods in the middle of the day. Efforts have been made to increase the use of electricity in commercial activities that will use energy during this time, while limiting demand at peak periods.

Energy options

Renewable energy options are increasingly well developed technically and markets are expanding, but severe constraints on the market remain. Governments of many developing countries are working to increase grid connectivity, although progress is slow and the growth often does not keep up with demand. The more densely populated areas tend to receive priority over others with lower populations, since more people can be connected to the grid supply for the same cost.

When attempting to bring renewable energy technologies within range of poor people it is important to work with the existing market to ensure that it is locally sustainable – not only economically and environmentally but in a range of factors that contribute to technological sustainability:

- local manufacture and product support;
- local ownership and management;
- community and individual financing, especially from micro-finance.

Involving rural people

Although there will be differences in the way projects are implemented, it has been found that successful projects involve the people affected in the planning and decision-making, often through the community committee. Many benefits are particular to women such as the provision of mechanized grain-milling services, replacing labour-intensive traditional methods of grain milling, and it is important to include women representatives in the committee.

Involvement of users results in a more efficient, rational use of resources and more equitable sharing of the benefits of development, and by involving users from the beginning the costs can be reduced by using local labour to build the infrastructure and, with training, carry out installations and maintenance thus ensuring a better and cheaper service for consumers.

Rural development is dependent upon making energy services more readily available to people living in remote areas. Ideally, energy services should be introduced within the framework of wider infrastructure and economic development. Combining development activities in such a way will strengthen the chances of successful community-based energy provision and enable the communities to improve their livelihoods and generate additional income.

Delivering electricity

Mini-grid system arrangements can include a distribution network with AC power stepped up to higher voltages for distribution, from 0.4 kV for lower-voltage distribution and from 11 kV for higher-voltage transmission lines. The electrification of villages with scattered houses and settlements using AC power systems requires a costly distribution network.

Hybrid systems combine renewable energy systems such as wind and solar with a diesel generator for a more consistent supply. There is a growing interest in integrated systems of energy delivery yet there is still only limited adoption of energy systems for the rural poor, primarily because hybrid energy systems add to the cost of energy delivery.

Stand-alone systems usually incorporate battery storage and have a 12 V DC circuit. The advantages of village electrification schemes using batteries include:

- The low-voltage battery excludes the danger of electric shocks within houses.
- The battery technology is a relatively simple and well known, being applied in vehicles throughout the world.
- The batteries make use of power during times of low load. This means that all available energy can be used even at times of low demand as the energy generated at this time will be used to charge the batteries.

The disadvantages include:

- The cost of electricity from rechargeable batteries can be very high.
- Battery life can be short if it is not properly used and maintained.
- They have a limited energy output which confines their use to lighting, radio and other small appliances which are not directly productive end-uses.

Another common practice for obtaining electricity is through pre-electrification battery-charging services. Remote energy systems cannot always supply power to all the households wishing to receive it because of limited capacity or because households cannot afford the tariffs or the connection charge. This has resulted in the establishment of battery-charging enterprises in which people can take batteries, usually lead-acid car batteries, to a centrally located energy supply such as a micro-hydro scheme. Battery charging can be done during the periods when the power system is not being used to its full capacity, thus improving the system's load factor.

Technologies for affordable electricity

One aspect to making energy schemes successful is to reduce the cost of the scheme through the various methods:

- sizing the system components to suit demand;
- local manufacture;
- selection of appropriate technology for components.

Small-scale manufacturing plays a huge role in the development of any region. Renewable energy technology can be used to stimulate indigenous manufacturing. The technology has to be appropriate for the region in question, or adapted to make it suitable. With the development of small-scale manufacturing, renewable energy can be introduced to more users at a lower cost than sophisticated imports, and equipment can be repaired and maintained more easily. Local manufacture creates employment and local added-value, improving the general economic situation.

Once a system is installed it has to be adequately maintained, so a support infrastructure needs to be established. This may require a training programme and appropriate documentation, regular refresher training, and an accessible supply of spare parts.

Transmission and distribution lines

Mini distribution systems require careful consideration as they can potentially add a huge amount of expense to a scheme. Standard distribution systems based on the principles of national grid systems are over-engineered for rural electrification schemes. Consequently, various low-cost alternatives have been used for such schemes.

Three-phase high-voltage. Three- or four-wire setups can be used for three-phase high-voltage systems. Four-wire systems use three-phase wires and a neutral return. The three-wire approach limits the voltage that can be supplied between phases, known as the phase voltage, acceptable for high-voltage (HV) distribution systems but not suitable for low-voltage systems. The main advantage is in the cost benefits associated with the reduction of the number of wires required.

Three-phase low-voltage lines. Three-phase wiring is relatively expensive for low-voltage distribution that can use single-phase options.

Single-phase low-voltage lines with wire return. There is a cost benefit in reducing the number of wires associated with single-phase systems compared to three-phase. The disadvantage of single-phase systems is that the power delivery is not as smooth as in a three-phase system, which can affect the performance of electrical devices.

Single wire earth return (SWER). A single-phase supply using the earth as the return reduces the costs even further by eliminating the return wire. The system was developed in New Zealand in the 1920s for rural energy supply.

In practice, a combination of transmission lines may be used depending on the size of the distribution grid in question. From the power house there could be a three-phase high-tension power line to minimize power losses, which can then be stepped to lower-voltage single-phase lines for local distribution. In most mini-grid systems the distance of the supply lines will only be a few kilometres. By comparison, national grid extensions to rural areas require much longer lines resulting in the need to upgrade the system to avoid excessive transmission losses.

Distribution lines need to be supported off the ground at a height that means they will not interfere with people's activities or transport, and will not be dangerous. The poles have certain requirements in terms of their size and strength, to counter wind conditions. In mini-grid systems the distribution poles can be a significant cost of the overall project.

In the home

Special approaches are required for low-cost electrification in the home if connections are to be economic. In subsistence farming communities, the average household expenditure on electricity can be less than $1 or £0.64 per month. The electricity consumption of low-income households is often just a few tens of kilowatt-hours (kWh) per month. The main problems faced by low-income households in obtaining an electricity supply are high initial connection charges and high costs of house-wiring.

With appropriate techniques, houses can be connected safely and with fewer dangers than those associated with the use of kerosene and candles. The dangers from electricity supply can be kept to a minimum by using earth-leakage circuit-breakers, flexible wiring systems, education, and regular safety checks. The high costs faced by new consumers can be reduced through the careful application of appropriate technologies such as prefabricated house-wiring systems, e.g. wiring harnesses and prefabricated distribution units (see below).

Load-limited supply

Load limiters have been successful in reducing the connection cost and the operating cost of electricity supply. The basic principle is to limit the current to a prescribed maximum. If the current exceeds the stated maximum then the limiter will disconnect the supply. The cost savings associated with load limiters are significant as they allow the reduction in scale of generation and transmission of electricity and in the time and cost of installation. Billing and revenue-collection costs can be reduced.

There are a number of options of load-limiting devices:

Miniature circuit breaker (MCB). This is the most common type of circuit breaker used and consequently the most familiar to electrical engineers. MCBs are mass-produced, robust and inexpensive.

Positive temperature co-efficient thermistors (PTCs). These devices are made from solid-state semiconductors. Again, the items are mass-produced and are used in consumer goods and telecommunication equipment. They are less common in household connections as they have a low current rating of typically 20–500 mA, which means they are not suitable for conventional electricity connections but can be used in restricted power supplies.

Electronic current cut-out. The electronic current cut-out (ECC) is a more recent option for limiting load. They were specifically developed in Nepal for this purpose. The ECC is not dependent on a change in temperature but measures the voltage. This voltage is an accurate measure of the load current and is used to turn off the electronic switch when the current is too high.

Prefabricated distribution units

The prefabricated distribution unit is known in South Africa as a ready board and in Papua New Guinea as a minimum service supply kit. It is a standard unit that is connected after the meter or load limiter and enables consumers to connect up their household with safety. In some cases they have a light fitting directly on top of them and in the cheapest form this may be the only load. Others have a number of breakouts for cables that can be used for additional loads. They incorporate consumer protection facilities including an earth-leaking circuit breaker, as well as over-current circuit breakers.

Wiring harnesses

Wiring harnesses are prefabricated and include the wiring as well as the distribution unit. They are complete house wiring systems that are quicker, easier and cheaper to install than the conventional approach to wiring houses. The harness is made to a standard format and available in a range of sizes so householders can determine the service level they require. Switches and light sockets are already built in at the time of assembly, and the wiring radiates out from the central control box that can include a load-limiting device and fuses.

The design was originally developed as a safe option for thatched houses, but has been widely applied to other forms of housing. Generally the wiring is not built into the walls of the house but is fixed to the wall surface. The cables can be quickly attached to the walls using self-locking cable ties. Any excessive length of wire is folded away rather than cut down so that fittings can be moved at a later date if required. This is particularly useful when extensions are added to a building. If local villages receive appropriate training, supported by the electricity supplier, then they are able to install wire harnesses.

Batteries

For renewable energy systems, it should ideally be possible to use most of the energy stored in a battery so that the time between recharging is as long as possible, but lead-acid vehicle batteries are the most readily available and most commonly used type of battery in renewable energy systems in developing countries. These batteries are designed to give a short burst of current to start the vehicle and then to be recharged immediately, so the depth of discharge is never very great. Consequently, the discharge should be kept within 30 per cent of the rated capacity and, in order to keep the battery in good condition and maintain its capacity and performance, it should not be left discharged for any length of time. Batteries may begin to fail after fewer than 100 cycles of discharging to 50 per cent of their capacity.

Where available, deep cycle or traction batteries are a better option as they can be discharged up to 80 per cent of their rated capacity with life cycles from 1000 to 2000. Batteries have been developed that are specifically designed for solar systems. They are delivered dry-charged and the electrolyte is added once they have been installed. The life-cycle range is typically around 1200 at 80 per cent discharge to 3000 at 50 per cent discharge. Sealed maintenance-free batteries have a good life cycle of 800 cycles at 80 per cent discharge, but they need to be recharged regularly to prevent sulphate build-up. These batteries are more expensive and less widely available but are more economical over their lifetime.

Lighting options

As lighting is usually the first use of domestic electricity systems in remote settings it is important to minimize the energy demand of the lighting units. Over recent years there has been a huge improvement in the efficiency of lighting units compared to traditional incandescent bulbs. See Chapter 15, 'Rural lighting' (Noble, 2012).

Low-wattage cookers

Cooking with electricity offers benefits to health and the environment, as it can replace fires that fill houses with smoke and cause many respiratory illnesses, and reduce the dependency on scarce resources of wood. Conventional electric cookers have a very high energy consumption, but low-energy electric cooking devices have been developed in Nepal by Development Consulting Services and are now manufactured commercially. Normal electric cookers consume about 1 kW per plate, which is far too high for the majority of renewable energy schemes. A simple meal for four people would need about 1 kWh of energy to cook, and generally people in a community tend to cook at about the same time. Low-wattage cookers can be left to cook food during the day when energy demand is low. They have high levels of insulation to keep the energy demand down.

In recent years health and environmental issues have become more prominent. Clean domestic energy reduces smoke exposure and lessens the need for fuel-wood thus reducing deforestation, land degradation and the consequent impact on climate change. Successful implementation of renewable energy schemes in rural areas is dependent upon a complex mixture of technological innovation combined with economical and institutional developments.

Low-wattage electrical cookers provide a clean environment and improve the load factor of micro-hydro schemes in Nepal

Source: Practical Action/Caroline Penn

Further information

Bibliography

Anderson, T., Doig, A., Rees, D. and Khennas, S. (1999) *Rural Energy Services: A Handbook for Sustainable Energy Development*, Practical Action Publishing, Rugby.

Mulugetta, Y, Doig, A., Dunnett, S., Jackson, T., Khennas, S. and Rai, K. (2005) *Energy for Rural Livelihoods: A Framework for Sustainable Decision Making*, Practical Action Publishing, Rugby.

Noble, Neil, ed. (2012) *A Handbook of Small-scale Energy Technologies: Practical Answers*, Practical Action Publishing, Rugby.

Smith, Nigel (1998) *Low-cost Electrification: Affordable Electricity Installation for Low-income Households in Developing Countries,* Practical Action Publishing, Rugby.

Ward, Stephen (2000) *A Guide to Producing Manuals and Facilitating Participation in the Planning of Off-grid Electrification Projects*, Practical Action Publishing, Rugby.

Wilkins, Gill (2002) *Technology Transfer for Renewable Energy: Overcoming Barriers in Developing Countries*, Earthscan and Routledge, Abingdon.

'Energy for rural communities' was originally written by Neil Noble in 2005. Neil has worked for Practical Action since 1998 as a technical adviser and as the Practical Answers coordinator. Previously he worked at the Rural Industries Innovation Centre in Botswana and studied Engineering Design and Appropriate Technology at the University of Warwick. He has a background in mechanical engineering within the UK.

Chapter 14
Refrigeration for developing countries

The options for refrigeration and cooling of foods are described in this chapter, from the simple passive evaporative cooler to the more sophisticated powered options. Technologies for energy provision are outlined and the issues relating to choosing a system are described.

Keywords: renewable, energy, refrigeration, food preservation, coolers, evaporative cooling

Introduction

Refrigeration plays an important role in developing countries, primarily for the preservation of food and medicine, and for air conditioning. Examples of these applications are listed below.

- *In agriculture and dairies:* removal of field heat immediately after harvesting of crops, storage of fruit, flowers, vegetables, milk and meat, and cooling during transport;
- *In retail trades:* sale of fresh foods, fish and cold drinks;
- *Buildings, computer installations:* air conditioning and temperature regulation;
- *Domestic:* food and drink storage;
- *Health clinics:* storage of blood, vaccines and medicine.

Choice of technology

Cooling can be provided in different ways. The method adopted in industrialized countries depends heavily on grid electricity, supplied continuously and reliably to every part of the country. In contrast, refrigeration is required in developing countries to stimulate agriculture and commerce, in vast areas that do not have a reliable electricity supply. Alternative methods are therefore necessary. A number of approaches can be considered, including three kinds of cooling technology that are contrasted in Figures 14.1–14.3:

- passive / evaporative;
- sorption heat-driven;
- mechanical compression.

The third method, mechanical compression, is usually dependent on a reliable and continuous supply of grid- or diesel-generated electricity. The other two

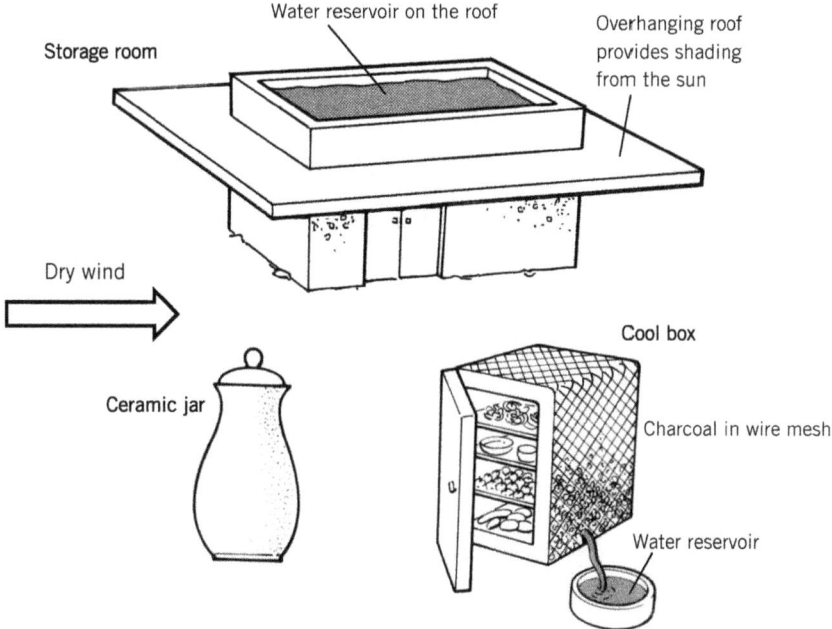

Figure 14.1 Passive/evaporative coolers
Source: Practical Action

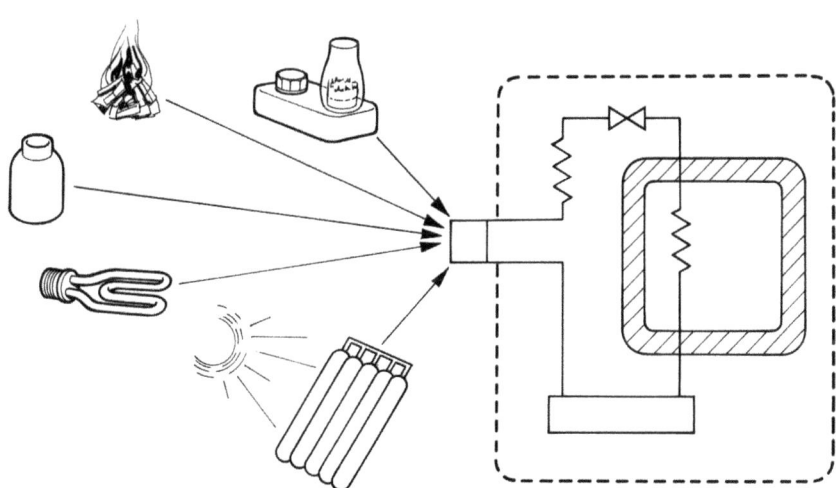

Figure 14.2 Sorption refrigerator
Source: Practical Action

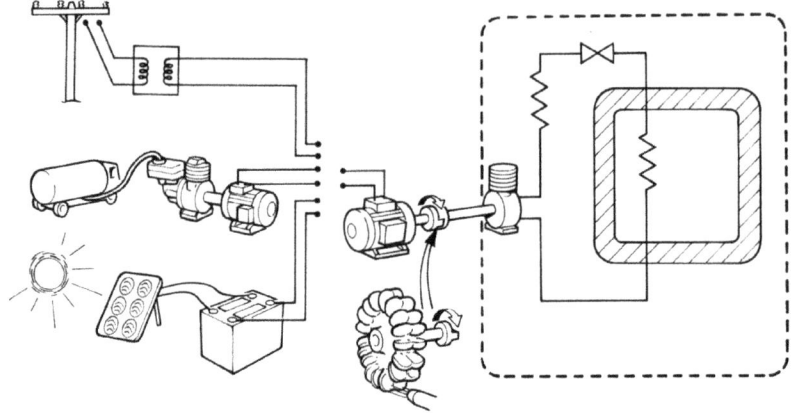

Figure 14.3 Compression refrigerator
Source: Practical Action

methods are therefore more suitable in non-industrialized areas. They require further development on the basis of requests from users in rural locations.

Several approaches that can be considered are as follows.

- Ice can be produced using electricity in regional centres, then transported to agricultural areas or used for packing of produce to be cooled in insulated containers. Electricity is either grid- or diesel-generated. Refrigerators that are electrically driven use mechanical compression technology.
- In some cases refrigerators can be driven directly by mechanical shaft power, for instance where water turbines can be readily installed.
- Production of ice using heat-driven coolers (HDCs) is possible on a local level at agricultural production points, allowing the packing of produce with ice for transport. The heat sources for HDCs are varied; heat can be from wood, charcoal or agro-waste burnt in open stoves, from fossil fuel in conventional burners, or from thermal solar collectors. HDCs use sorption technology.
- Provision of cold storage chambers using passive, sorption, or mechanical compression technology. If passive cooling is used, temperatures lower than 10°C can rarely be achieved.
- Provision of cold storage at the point of use using mechanical compression coolers drawing electricity from photovoltaic cells. This is referred to as photovoltaic cooling technology.

The most suitable method of cooling chosen will depend upon various factors: the application, the degree of reliability required, the supply of power, the level of skill needed to operate and maintain, training facilities, and available finance. The different technologies should be considered with these factors in mind. As with any technology, sufficient training is especially important; it must be planned as an integral part of an implementation programme and remains a constant concern during the years following installation. This will increase reliability of the system and reduce life-cycle costs dramatically.

Temperatures and ventilation

Different applications have different requirements for temperature control and ventilation. Figure 14.4 shows the temperatures needed for the storage of butter, meat, fresh fish, and milk. Very often storage of vegetables is complicated by the need for careful ventilation to remove unwanted gases, and to avoid humidity conditions, which would spoil the produce. Relative humidity requirements vary depending on the moisture content of the produce. A simple method of increasing humidity is to sprinkle water on the floor. In vaccine and blood storage very careful temperature control is required.

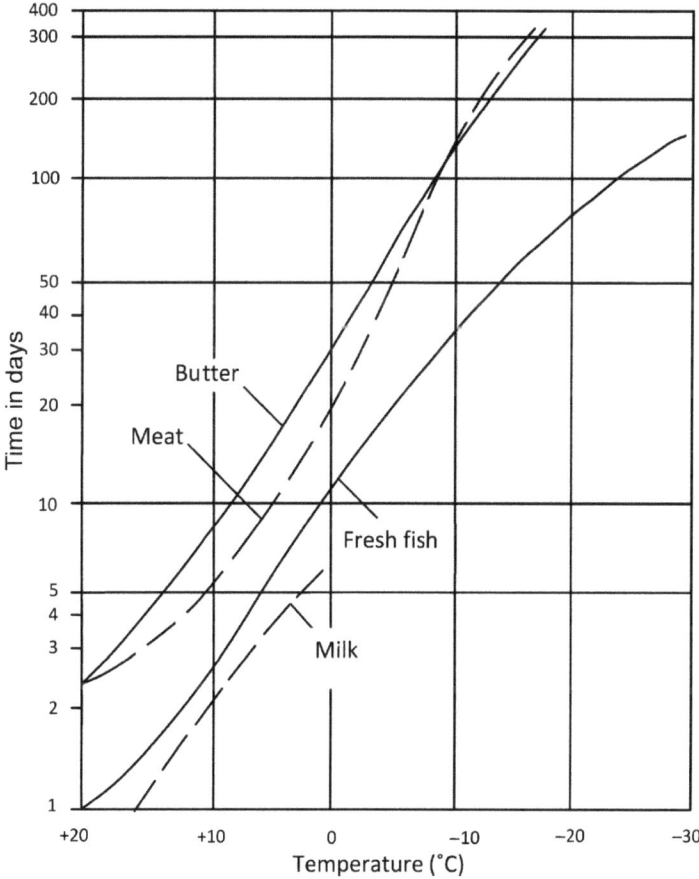

Figure 14.4 Temperatures for safe storage

Source: Practical Action

Passive / evaporative cooling

In applications where temperatures of 10–25°C are needed, passive methods can be used; see Practical Action Technical Brief, Evaporative Cooling. These include traditional methods such as the use of porous jars or wet sack coverings, where the evaporative heat of the liquid, usually water, is drawn into the atmosphere. This method is effective where the atmosphere is naturally dry. Domestic storage devices have been designed along these lines, particularly with the use of charcoal beds, drip-fed with water.

Nocturnal cooling can be effective in areas where clear night skies are common. In air-conditioning applications, the use of shade has been developed effectively in traditional architecture, together with evaporative cooling by fountains and roof ponds.

Wherever possible, passive methods should be used both in agriculture and architecture, since they can be sustained locally and are economic. Only when cooling below 10°C is needed is it justifiable to look at active cooling technology, requiring complex machinery, and technical maintenance programmes.

Sorption heat-driven coolers (HDCs)

The principle of sorption refrigeration is shown in Figure 14.5 which illustrates the simplest type of sorption cooler and which has an intermittent cycle consisting of two phases. Continuous cycles are also possible – the Electrolux uses a continuous cycle. The general term sorption covers both liquid absorption and solid adsorption variants of this technology. Sorption units have some very important advantages. They can be designed to contain no moving parts, so that skilled maintenance personnel and replacement of components are less likely to be needed. Secondly, they are simple to manufacture; local manufacture increases local knowledge of the technology, which improves operation, maintenance and fault finding. Thirdly, they are readily adaptable to locally available fuels, including biomass and solar energy. Finally, the refrigeration circuit does not use CFCs, which damage the environment. Sorption units are referred to as HDCs (heat-driven coolers).

Conventional gas- and kerosene-driven sorption units

The heat source in conventional sorption refrigerators is usually a gas or kerosene flame. Units powered from gas bottles are used on caravans or boats. A domestic unit, often used in remote locations in developing countries, is the kerosene-driven Electrolux. It has been calculated that the cost of purchasing and running one of these units is around £1,000 ($1,562 at a conversion rate of $1.00 : £0.64) for 10 years' use. The refrigeration circuits of these devices operate reliably for many years. Maintenance of the burner assembly is required and a constant supply of wicks, burners and lamp glasses is essential. Lastly, the fuel tank must be replenished with kerosene of suitable quality. These units involve

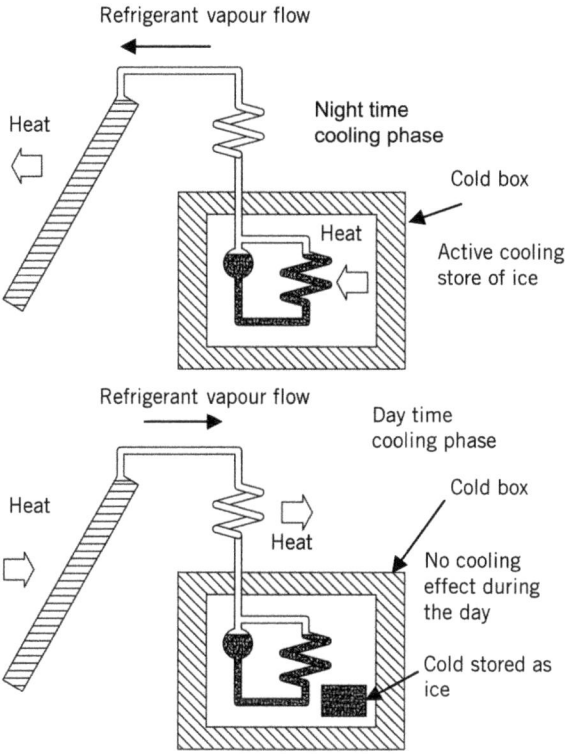

Heat phase: At high bed temperature and systems pressure, the refrigerant is driven out of the sorbent and collects in the cold box as a liquid

Figure 14.5 Sorption cooler

the use of hydrogen as a working fluid and cannot be designed as efficient ice-makers, although they have some ice-making capacity.

Novel sorption units

Novel sorption units are under development for greater efficiency in ice making and cold storage. They do not involve hydrogen as a working fluid. A great deal of emphasis is being placed on design for reliable operation in remote environments where technical maintenance services are not available. Emphasis is also placed, in some cases, on design for local manufacture. Costs and performance figures are not easily available since these units are still on trial.

Solar-powered sorption units

The heat source for sorption units of the kind shown in Figure 14.5 can be the sun. In a simple version the heating phase ends at sunset, and the refrigeration phase occurs during the night. If the sun fails to shine for a few days, the ice made on previous days acts as a store of cold, keeping the cold box at a low temperature while it gradually melts. It is expected that a unit producing 100 kg of ice per day can be produced for £4,000 or $6,250 (including the cost of highly efficient solar thermal panels), giving an ice cost of £0.03 per kg.

A project that Practical Action is working on in collaboration with SCORE at the University of Nottingham is a thermoacoustic device that uses thermal energy from a stove or solar energy to generate acoustic waves that then produce electricity. More information is available at www.score.uk.com.

Mechanical compression

Grid electricity

Where a reliable electricity supply exists, the most economic option is to install a standard compressor driven unit. Conventional refrigerators of this kind are sold commercially. As an example, a unit making about 100 kg of flaked ice, for fisheries use, each day in tropical conditions will cost £7,000, not including the cost of storage containers for the ice, or delivery. The power consumption would be in the order of 4 kW continuously. There will be extra costs in the form of replacement parts, maintenance and ancillary equipment.

Water-turbine-driven coolers

Costs can be reduced if shaft power is used directly to drive the compressor, for example from a water turbine. An auxiliary electricity supply is useful to provide control and protection functions, and for instance to drive ventilation fans. It is nevertheless feasible to design wholly mechanical cold storage and ice-making systems.

Diesel generating sets

The cost of operating a generator in rural areas is dependent on local conditions and must be assessed in the light of local experience. Quite often the cost can be very much higher than expected because of the need for maintenance personnel and the difficulties encountered in obtaining fuel and spare parts. If the generated electricity is not available continuously then the refrigerator should be designed as an ice-maker, allowing cold to be stored in the form of ice. Experience has shown that systems involving the storage of electricity in batteries have very high costs and are unreliable.

Solar photovoltaic systems

Solar energy is an intermittent power source, usually available for 12 hours every day in sunny climates. The intensity of insolation is very variable. It can be converted by photovoltaic cells into electricity, which is then stored in batteries, so that a continuous smooth electrical supply can be provided to power a mechanical compression refrigerator. See also Practical Action Technical Brief, Solar Photovoltaic Refrigeration of Vaccines.

The advantage of using solar power is that it is a source that can be relied upon never to fail for more than a few days. This reliability is very important in some cases, such as vaccine storage, where loss of temperature control can spoil the vaccines completely. The battery is designed to continue to provide electricity at night and on days when no sunshine is available. In this application, the high cost of photovoltaic cells, batteries and control equipment is justified. The size of the photovoltaic array and the battery capacity must be carefully calculated to provide an economic system.

Solar refrigeration units of this kind, especially designed for vaccine preservation, are commercially available. A system providing 60–80 watts of cooling is typically priced in the range of £3,000–5,500. Replacement parts will tend to cost £500–1,000 in the course of four years of operation. Most of this cost will be in the replacement of batteries which are designed to have a four-year life but can fail in a shorter period if maintained poorly. Replacement costs are considerably reduced if skilled technical maintenance personnel are available.

Combined heating/drying/cooling system

Because a refrigerator releases heat it can be used to raise temperatures in agricultural processes like crop or spice drying. The cooling effect can be used to dehumidify the air passing over the crop and the heating effect can be used to warm the air. In this, very high efficiencies can be obtained (for instance up to seven times as much useful energy produced as required to drive the device). Such efficiencies are commonly met in timber-drying plants using these principles. Practical Action developed low-cost methods of utilizing this effect, with respect to drives from small hydro turbines or from steam or diesel engines.

A second example is the use of heat from a refrigerator (also known as a heat pump, exactly the same machine) to help sterilize milk, while the same refrigerator cools the milk to preserve it.

Choosing the system

In order to decide which refrigeration system to adopt for a particular purpose, it is necessary to consider the ongoing inputs required by each system. Table 14.1 lists the various systems and the inputs required for each. The choice of system is based on the foreknowledge that all the necessary inputs will continue to be available in the locality of the refrigerator. The mistake is often made of installing a unit with a relatively low purchase cost which later ceases to function through lack of necessary inputs.

Table 14.1 Comparing refrigeration systems

System	Comment	Energy source	Personnel	Spare parts	10-year total cost
Mechanical compression					
Grid	Electricity available for other purposes, e.g. lighting, flaking of ice	Grid electricity. Cost of connecting/transforming can be high	Maintenance: skilled personnel	Source of parts may be distant. Supply may be uncertain	Purchase cost, electricity, personnel, replacement parts
Diesel	Electricity available for other purposes, e.g. lighting, flaking of ice	Diesel generator	Maintenance: skilled personnel permanently on-site	Source of parts may be distant. Supply may be uncertain	Purchase cost, diesel, replacement parts
Solar photovoltaic	Expensive. Electricity available for lighting, communications, temperature control	Irradiation of 10–20 MJ/day/m². Long cloudy periods problematic	Skilled personnel permanently available	Battery life 2–4 years. Control electronics can fail	£3,500–6,500 for 60–80 W cooling, includes replacement costs
Sorption (HDCs)					
Conventional	Well known in the field	Gas/kerosene quality must be adequate	General maintenance of burner parts, wick adjustment, etc. Local skills sufficient	Replacement of burner parts routine	£1,000–2,000 for 60–100 W cooling and small maintenance cost
Novel (biomass driven)	Local manufacture possible. Under development, relatively low cost	Any locally available heat source suitable, eg charcoal, coal, agro wastes, cow dung, fossil fuels	Maintenance of open burner, brine tank, cooling water. Local skills sufficient	Locally available spare parts	Purchase cost projected at £2,000 for 100 kg ice/day. Fuel cost £50–100 per year
Solar	Very new on market. Performance not yet evaluated. Low night temperatures can be advantageous in some cases. Presence of cooling water advantageous. Provincial manufacture possible	Solar irradiation 10–120 MJ/day/m². Long periods of cloud problematic	Local skills sufficient, few moving parts	Solar panels may require import of spare parts	Current purchase cost £1,500–2,500 for 10 kg ice/day. Projected cost £4,000 for 100 kg ice/day (including solar panel)

Note: £1 = US$1.56 as at August 2012

Further information

Technical Briefs and other fact sheets

Evaporative Cooling, http://practicalaction.org/evaporative-cooling-1
Evaporative Cooler – The Ceramic Refrigerator, Engineering drawings of the
   Sudan Zeer pot-in-a-pot evaporative cooler.
   http://practicalaction.org/evaporative-cooling-the-ceramic-refrigerator-1
Solar Photovoltaic Refrigeration of Vaccines,
   http://practicalaction.org/solar-photovoltaic-refrigeration-of-vaccines-1
Cold Storage of Fruit and Vegetables,
   http://practicalaction.org/cold-storage-of-fruit-and-vegetables-1
Solar Energy and Rural Health Care: Fact Sheet N132:
   https://apps.who.int/inf-fs/en/fact132.html WHO
Solar Energy for Cooling and Refrigeration,
   www2.warwick.ac.uk/fac/sci/eng/staff/dbm/es368/solarcool.pdf DTU

Bibliography

American Society of Heating, Refrigeration and Air Conditioning Engineers
   (ASHRAE) (1997) *Fundamentals,* ASHRAE, Atlanta.
ASHRAE (1998) *Refrigeration,* ASHRAE, Atlanta.
ASHRAE (1999) *HVAC Applications,* ASHRAE, Atlanta.
ASHRAE (2000) *HVAC Systems and Equipment,* ASHRAE, Atlanta.
Dossat, R.J. (1997) *Principles of Refrigeration.*, 4th edn, New Jersey, Prentice Hall.
Tomkins, Ray (1985) *Prospects for Solar Refrigeration,* unpublished manuscript.
USAID (n.d.) *Powering Health* [website] www.poweringhealth.org/topics/refrigeration/
   index.shtml [accessed 4 September 2012].

Useful addresses

American Society of Heating, Refrigeration and Air Conditioning Engineers (ASHRAE)
1791 Tullie Circle, North East Atlanta, GA 30329, USA
Tel: +1 (404) 636 8400, Web: www.ashrae.org

EPI (Expanded Programme on Immunization), World Health Organization
Avenue Appia 20, 1211 Geneva 27, Switzerland
Tel: +41 (0)22 791 4517, Web: www.who.int

Institute of Food Research
Norwich Research Park, Colney, Norwich, NR4 7UA, UK
Tel: +44 (0)1603 255 000, Web: www.ifr.ac.uk

Institute of Refrigeration
Kelvin House, 76 Mill Lane, Carshalton, Surrey, SM5 2JR, UK
Tel: +44 (0)20 8647 7033, Web: www.ior.org.uk

International Institute of Refrigeration
177 Boulevard Malesherbes, 75017 Paris, France
Tel: +33 (0)1 4227 3235, Web: www.iifiir.org

Ray Tomkins Management School
Imperial College, 53 Princes Gate, Exhibition Road, London, SW7 2PG, UK
Tel: +44 (0)20 7594 9137, Web: www.ms.ic.ac.uk

Warwick Energy Research Group
School of Engineering, University of Warwick, Coventry CV4 7AL, UK
Tel: +44 (0)24 765 23137
Web: www2.warwick.ac.uk/fac/sci/eng/research/energy

Manufacturers and suppliers

Sibir International AB
Torggatan 8, S-171 54 Solna, Sweden
Tel: +46 (8) 501 025 08, Web: www.sibir.com
Domestic kerosene refrigerators, domestic gas refrigerators, medical kerosene refrigerators, medical gas refrigerators

Total Refrigeration Ltd
Unit 2A East Tame Business Park, Rexcine Way, Talbot Road, Hyde, Cheshire SK14 4GX, UK
Tel: 0845 127 2527, Web: www.totalrefrigeration.co.uk
Icemakers, cabinets, chillers and freezer cold rooms

Ziegra Ice Machines (UK) Ltd
Unit 2, Phoenix Court, Hammond Avenue, Stockport, Cheshire, SK4 1PQ, UK
Tel: +44 (0)161 429 0525, Web: www.ziegra.co.uk
Ice machines and ice storage systems

'Refrigeration for developing countries' was last updated by Neil Noble in June 2012. Neil has worked for Practical Action since 1998 as a technical adviser and as the Practical Answers coordinator. Previously he worked at the Rural Industries Innovation Centre in Botswana and studied Engineering Design and Appropriate Technology at the University of Warwick. He has a background in mechanical engineering in the UK.

Chapter 15
Rural lighting

The alternatives to mains electricity are covered in this chapter, including the methods of providing light and the technologies involved, from the simple candle to the white light-emitting diode. Much of the information is focused on stand-alone electricity supply, such as solar lanterns.

Keywords: renewable, energy, rural lighting, lighting technologies, light-emitting diodes, solar energy

Introduction

Lighting is taken for granted in industrial countries. It is hard for many people to imagine life without being able to obtain light at the flick of a switch, but grid electricity does not extend to many rural areas in developing countries and is not likely to be available in the near future. Even in urban areas many people have difficulty accessing clean energy sources as they are not connected to the grid or have unreliable power supplies.

Lighting in the rural areas of developing countries is often provided by candles or kerosene lamps. To a lesser extent, biogas, diesel generators and renewable energy systems are used. Torches (or flashlights) powered by expensive, disposable dry-cell batteries are used as a portable source of light for intermittent use. Access to modern energy products is hindered by poor market-supply chains, low awareness of existing technologies and lack of access to finance for consumers and entrepreneurs.

What is light?

Light is electromagnetic radiation; the human eye is sensitive to a spectrum with visible colours as seen in a rainbow. When these colours are mixed they appear as white light. More energy is present in the violet/blue end of the spectrum than at the red end. Therefore, more energy is generally necessary for the blue-violet component needed to produce what to the eye appears as white light. So the quality of light (in colour terms) influences the energy requirement; if colour does not matter then it is at least theoretically more efficient to use a red or orange light, and this in fact is common practice in the case of street lighting where energy-efficient lights are used, namely orange/yellow sodium lights. For domestic purposes people prefer to pay extra to get white or near-white light.

Light intensity, or illuminating power of a light source in any one direction, is commonly defined in candela, which although it has a rigorous scientific definition, for practical purposes can be thought of as *candle-power*; i.e. the output from a standard paraffin-wax candle. The rate at which light is emitted is measured in lumens, which are defined as the rate of flow of light from a light source of one candela through a solid angle of one steradian. A more easily understood approximation of this would be to imagine a one candela candle at the apex of a conical lampshade with its sides sloping at about 70 degrees to each other; the conical beam emitted, diverging at about 70 degrees and lighting an area of $1m^2$, would be about one lumen. Lamps are often rated in lumens. Another unit measuring lighting is the lux, which is used where the area to be lit is accounted for. When measuring the light intensity in a room rather than the power of a lamp the lux can be used.

Methods of providing light

There are three main physical principles by which light may be produced:

- incandescent;
- fluorescent;
- electroluminescence.

Incandescent light

The incandescent principle depends on heating a source to a temperature in the region of 2000, 4000 or 6000°C to obtain reddish, yellowish or white light, respectively. Typical examples are candles and lamps utilizing a bright flame (where white-hot or incandescent particles of carbon in the flame produce the light) and incandescent filament light bulbs where a fine coil of tungsten wire is heated (in a vacuum or inert, low-pressure gas to prevent the filament oxidizing or burning) by an electric current passing through it. All incandescent light sources, whether flames or electrically heated filaments, tend to produce more heat than light and are therefore relatively inefficient in the rate of conversion of energy to light.

Fluorescent light

Fluorescence occurs under certain conditions when a material can be made to glow with a 'cold light'. Fluorescence is a phenomenon in which the atoms of a gas, vapour or solid are excited in such a way that they emit light. In some cases, such as sodium and mercury vapour discharge lamps (used commonly for street lighting), vapour in a glass tube emits the light. In other cases, such as the commonly used fluorescent tube lights, ultraviolet light, which is invisible to the eye, is emitted by exciting mercury vapour atoms within the tube, and this in turn causes a white translucent coating in the tube to fluoresce with a whitish light. In other words, the coating converts invisible ultraviolet light into visible white (or near-white) light. Most fluorescent processes involve some expenditure of energy,

A biogas powered lamp in Bandarawella, Sri Lanka. Biogas is piped directly into the home and used for cooking and lighting

Source: Practical Action / Zul

so they are accompanied by the production of some heat. (Non-electrical fluorescence is used where the mantles of pressure lamps and gas lamps are heated and emit a much brighter and whiter light than would occur simply as a result of their temperature.)

Electroluminescent light

White-light-emitting diodes (WLEDs) produce electroluminescence when electrons recombine with holes releasing photons. The colour of the light depends on the energy the photons have which is determined by the jump the electrons make in the semiconductor.

The advantages of WLEDs over other lighting sources include their very low energy requirements, their long working life, robustness and reliability. WLEDs are becoming more popular because of their low energy demand, although their availability is still limited in many regions and they are relatively costly. Currently high-performance WLEDs cost roughly twice as much as regular LEDs and more than incandescent bulbs. Although WLEDs consume very little energy compared to an incandescent bulb, luminous efficiency (lumens per watt) can vary considerably between different products. There can also be a high degree of variation within individual batches of products.

136 HANDBOOK OF SMALL-SCALE ENERGY TECHNOLOGIES

Testing a solar WLED lantern in Nepal
Source: Practical Action / Rakesh Shrestha

The number and configuration of the diodes will vary between different lamps. The light from WLEDs is very directional which is not always best suited for general lighting in a room; reflective lampshades can improve the situation.

One of Practical Action's projects in Nepal promoted WLED lamps incorporating three diodes and rechargeable batteries that were made by trained local community members in partnership with Krishna Grill & Engineering, a local manufacturing company. Light up the World, another organization that promotes lighting to poor rural communities in Nepal, used simple hand-operated generators to power the low-energy lighting systems in remote parts of the country.

For practical purposes, the options for lighting reduce generally to either lamps that run on fuel, or electric lights. Table 15.1 indicates the options and their relative lighting capability.

More important to the user than the efficiency in lumen/watt is the cost per lumen. However, this is difficult to determine as prices will vary depending on where items are purchased and on the quantity bought, as well as the actual life of the item.

Lighting options that use a wick, such as kerosene lamps and candle wax, are often the cheapest sources of watts which partially makes up for their inefficiency, but not for their poor quality light. Kerosene pressure lamps produce better light, but they are unpleasantly noisy and uncomfortably hot to be near in a tropical climate; they also use much more fuel than wick lamps and are troublesome to

Table 15.1 Lighting capacity

Type of light	Energy source	Intensity (lumens)	Efficiency (lumen/W)
Candle	Paraffin wax	1	0.01
Oil lamp (wick)	Kerosene	1–10	0.01–0.1
Hurricane lamp (wick)	Kerosene	10–100	0.1–0.2
Oil lamp (mantle)	Kerosene	1000	1
Gas lamp (mantle)	LPG (e.g. butane or biogas)	1000	1
Filament lamp 3 W	Electricity	10	3
Filament lamp 40 W	Electricity	400	10
Filament lamp 100 W	Electricity	1300	13
Fluorescent 15 W	Electricity	600	40
Fluorescent 30 W	Electricity	1500	50
Mercury 80 W	Electricity	3200	40
Sodium SOX 35 W	Electricity	4500	128
White light-emitting diode 1 W	Electricity	25–50	25–50

start. Cylinder gas lights provide a slightly more expensive but a more convenient alternative.

All combustion lamps and candles pose a real fire risk and contribute to indoor air pollution and its related health problems. They also emit greenhouse gases which increase global temperatures and change weather patterns. Kerosene and butane are expensive and sometimes in short supply in developing countries. Not only does fuel need to be bought but it may also mean a long journey to the market to acquire it. Therefore the rest of this chapter addresses the use of electric lighting in rural 'off-grid' locations.

Solar photovoltaic lanterns

Solar lanterns are an all-in-one system incorporating the battery storage and controls along with the lighting unit. They are easy to use and do not require any installation. There are many solar powered lanterns available. Practical Action (then ITDG) formed a partnership with Sollatek, a company that specializes in solar and electronics manufacture, to develop two versions of a robust solar lantern called the Glowstar and the Glowstar plus. Costs for a solar-powered lantern can range from less than £10 to more than £100 with varying performance.

Home electrical systems

Low-cost systems that require a small amount of electricity usually from a renewable energy source and using fluorescent lighting (or increasingly WLEDs) can be installed into many locations. They are sometimes provided in kit form and are linked to a mini grid or are linked to their own energy supply such as a wind turbine or solar panel when there is a reasonable source of power. Correct assessment and placement of solar panels and wind turbines are important in making these technologies work efficiently.

Batteries

Energy generated by a solar panel or wind turbine needs to be stored until it is required and this is done using batteries. Solar lanterns have an inbuilt rechargeable battery. It is possible to use disposable batteries (i.e. dry cells) which are bought ready charged and thrown away when exhausted. These are convenient to use but extremely expensive in terms of electrical energy costs and need to be transported and distributed to rural areas raising the cost even more and making the supply unreliable. There is also an issue of waste disposal once they are finished with. Rechargeable batteries are more cost-effective than disposable batteries, but recharge equipment then becomes necessary. The two main options available are:

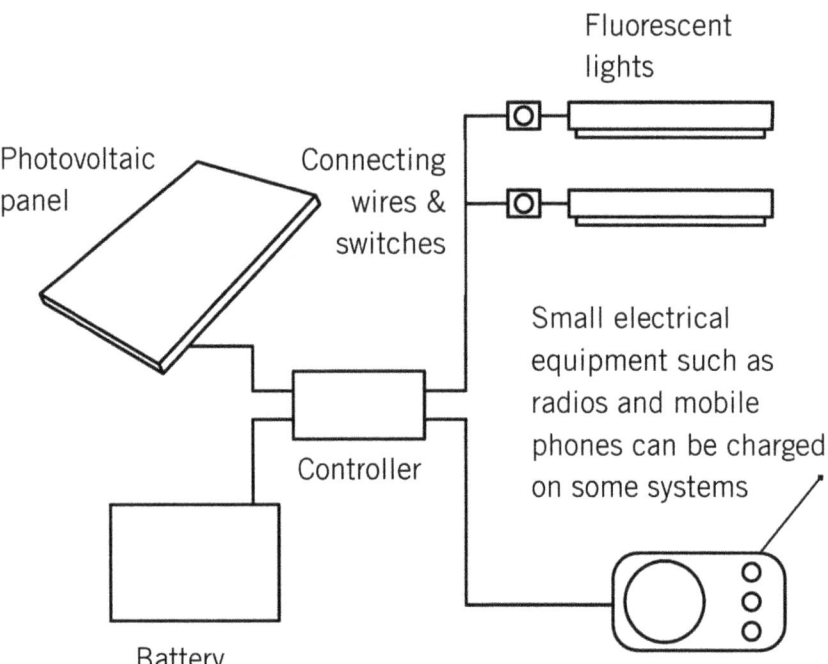

Figure 15.1 A simple solar system for lighting

Source: Practical Action

- nickel-cadmium batteries;
- lead-acid batteries.

Nickel-cadmium batteries are generally less widely available (except as dry-cell substitutes) and cost more, but they can be more robust and tolerant of abuse than lead-acid batteries. However, they self-discharge quite quickly if not used. Electrical energy from a lead-acid battery can cost as little as one twentieth to one fifth as much for the same amount of energy delivered from primary (dry) batteries.

Lead-acid batteries are similar to the batteries used in cars and in many cases car batteries are used in lighting systems as these are the most widely available type of battery. For most lighting purposes these are the easiest and cheapest option. Lead-acid batteries are also available as deep-discharge batteries which have a longer life than car batteries, and if looked after tend to be better for general electrical storage. An important point to note with lead-acid batteries is that their life is considerably shortened if they are over-discharged. Ideally they should only be discharged to about 30 per cent of their full rating; i.e. a 60 Ah (ampere-hour) battery should only be discharged to 30 Ah before recharging it.

Most lead-acid and nickel-cadmium batteries require regular checking of their electrolyte level and topping up with distilled or deionized water (not with acid). Rainwater can be used for this purpose, providing it has not been contaminated in any way. Low-maintenance and maintenance-free lead-acid batteries are also available, at slightly increased cost.

Typically deep-discharge lead-acid battery costs are in the region of £60/kWh of total rated capacity or £120/kWh of usable storage, while nickel-cadmium will be about three times this level. The cost increases with some specialist types and very small batteries. Batteries are generally provided with nominal voltages in multiples of 2 V; common larger capacity lead-acid batteries will be 12 V or 24 V nominal voltage.

Cable runs should be kept as short as possible with low-voltage supplies (or heavier cable must be used), otherwise significant losses will occur in the cables. The voltage drop is numerically equal to the current in amps multiplied by the resistance of the cable in ohms. A 10 m length of 0.75 mm² cable with a specific electrical resistance of 0.042 ohms carrying 1.25 amps will have a voltage drop of:

$$1.25 \times 0.042 \times 10 = 0.525 \text{ V}$$

This represents about 96 per cent cable efficiency, which is acceptable. However, 100 m of the same cable will cause a voltage drop of 5.25 V, which will cause the light not to work and in any case represents a quite unacceptable loss of nearly 50 per cent of the energy supplied.

A 30 W fluorescent light (for example) running off a battery with an inverter for six hours per night will consume 180 Wh each 24 hours. Losses in the inverter, cables and the battery will increase this requirement to about 300 Wh/24h. To avoid more than 50 per cent discharge and provide a nominal 24 hours' storage capacity would require, with the above example, a battery with a usable capacity of 600 Wh (1200 Wh to total discharge), costing in the region of £80.

The energy sources

There are four main methods by which the electricity for battery charging may be provided:

- taking the battery to the nearest mains supply and putting it on charge;
- using a small engine-powered generating set;
- using renewable energy such as photovoltaic or wind charging systems;
- using a hybrid system.

Using a generating set imposes considerable problems as, unless power is being generated for other purposes too, the charging current which is acceptable for small battery storage for just one or two lights is rather low for even the smallest generating sets; hence the engine needs to be run at part load which results in inefficient fuel use and is bad for the engine.

A photovoltaic system is often the most applicable stand-alone option, as adequate sun for charging can be found in most parts of the world, and such a system, apart from occasional cleaning of the array, requires little attention. However, solar photovoltaic arrays are still expensive. Current costs (delivered and installed) are in the region of £3–5/Wp (peak watt); the supplier should be able to advise on the size needed (and can generally supply a battery and lighting system too).

To run a 30-watt light for six hours would typically require, in a sunny tropical location, two nominal 40 W solar modules, and would therefore cost in the region of £400. Areas with extended cloudy periods may need up to twice this capacity.

Another option is to use a small wind generator. This can be cheaper than solar power in locations with mean wind speeds above 4.0 m/s in the least windy months. Some of the WLED lamps used in Practical Action's Nepal project were charged by small wind turbines rated at 200 W, supplied by Krishna Grill & Engineering. A separate Technical Brief, Wind for Electricity Generation, gives further information on wind turbines and gives a rule of thumb that the required rotor area of the wind generator for small-scale applications like lighting will be:

$$\text{Rotor area (m}^2\text{)} = \frac{\text{Energy demand (kWh/day)}}{0.0048\ V^3}$$

where V is the mean wind speed in the least windy month of the year. For the 30-watt lamp for six hours example, it follows that a 0.5 m² rotor wind generator will suffice at 5 m/s, a 1 m² rotor is needed at 4 m/s and a 2.3 m³ rotor at 3 m/s. The typical cost of wind generators is around £400–600/m² for small machines (installed).

In our example of a 30-watt fluorescent light, which is comparable in illuminating power to a 100-watt tungsten filament lamp and brighter than a pressure kerosene lantern with a mantle, in a rural 'off-grid' location using wind or solar power, the following items would be needed:

Item	Approximate cost
12 V 30 W fluorescent light with built-in inverter unit	£10
10 m cable, connectors, etc.	£10
1200 Wh battery (rated) giving 6000 Wh usable capacity	£80
plus either:	
60–90 W (rated) solar array with battery charge regulator (depending on solar irradiation)	£200–500
or	
0.5 to 2 m² rotor area wind generator with battery charge regulator	£200–900

This may sound a lot to pay for up to five lights, but the running costs will be negligible and good quality light will be reliably available. A pressure lantern, which is much less satisfactory, would cost only about £15–30, but it would consume in the region of 5 litres of kerosene per week on the same duty cycle, which would cost typically £75–150 per year.

Although the price of solar panels and small wind generators is still high they are declining in real terms as production increases, markets expand and awareness increases while the price of kerosene will continue to rise. There are tens of thousands of small photovoltaic lighting systems in use in developing countries and tens of thousands of wind-powered lighting systems, particularly in China.

To overcome the difficulties associated with the high initial costs of renewable energy systems for low-income households in developing countries, financing schemes are often set up where the user can pay for the equipment in instalments or rent the equipment off the supplier. Grameen Shakti has managed to install 300,000 solar home systems in Bangladesh using small loan schemes.

Further information

Technical Briefs and other fact sheets

Batteries, http://practicalaction.org/batteries
Candle Making, http://practicalaction.org/candle-making
Kerosene and Liquid Petroleum Gas (LPG),
    http://practicalaction.org/kerosene-and-liquid-petroleum-gas-lpg
Lighting Africa, Market research,
    www.lightingafrica.org/resource/market-research.html
Lighting Africa, Technical research,
    www.lightingafrica.org/resource/technical-research.html

Lighting Africa, Related resources from outside organizations,
   www.lightingafrica.org/resource/resources-other-organizations.html
Solar Energy: A Reference Guide for Users,
   http://practicalaction.org/solar-energy-a-reference-guide-for-users
Solar Lanterns Tests, GTZ (now GIZ),
   www.lightingafrica.org/files/1938-09_GTZ_Solarleuchten_engl02.pdf
The Lumina Project, Technology assessment,
   http://light.lbl.gov/technology-assessment.html
The Lumina Project, Best-practice guide for off-grid lighting product development,
   http://light.lbl.gov/best-practices.html
Wind for Electricity Generation,
   http://practicalaction.org/wind-for-electricity-generation

Bibliography

Louineau, Jean-Paul, Dicko, Modibo, Fraenkel, Peter, Barlow, Roy and Bokalders, Varis (1994) *Rural Lighting: A Guide for Development Workers*, Practical Action Publishing, Rugby.

Useful contacts

Centre for Renewable Energy
PO Box 589, Kathmandu, Nepal
Tel: +977 1 248852/351052, Web: www.crenepal.org.np
Promoting WLEDs in Nepal

Grameen Shakti
Grameen Bank Bhaban, Mirpur-2, Dhaka-1216, Bangladesh
Tel: +8802 9004081, 900431, Web: www.gshakti.org
Solar home systems for Bangladesh

Light up the World
University of Calgary, Mechanical Engineering, 1111 22nd Ave NW, Calgary,
   Alberta T2M 1P6, Canada
Tel: +1 (403) 284 2596, Web: www.lutw.org

Lights for Learning
69 High Street, Cricklade, Wiltshire, SN6 6DA, UK
Tel: +44 (0)1793 75084, Web: www.lightsforlearning.org
LED lighting systems for educational projects in Africa

Lighting Africa
PO Box 30577-00100, Nairobi, Kenya
Tel: +254 20 275 92 00, Web: www.lightingafrica.org
A joint IFC and World Bank programme, Lighting Africa is helping develop commercial off-grid lighting markets in sub-Saharan Africa as part of the World Bank Group's wider efforts to improve access to energy.

Solar Electric Light Fund (SELF)
1775 K Street, NW Suite 595, Washington, DC 20006, USA
Tel: +1 202-234-7265, Web: www.self.org
A non-profit charitable organization. SELF seeks to help communities and governments in the acquisition, financing and installation of decentralized household solar electric systems in the developing world.

SolarAid
Unit 2, Third Floor, Pride Court, 80-82 White Lion Street, London N1 9PF, UK
Tel: +44(0)20 7278 0400, Web: http://solar-aid.org/
Developed a solar LED lantern kit to fit kerosene lamps. This project is focused on Malawi.

Winrock International India
7, Poorvi Marg, Vasant Vihar, New Delhi 110 057, India
Tel: +91 11 614 2965, Web: www.winrockindia.org
Developed a solar lantern for India

Equipment suppliers

d.light
Hong Kong
Tel: +852 3106 6300, Web: http://dlightdesign.com/home_global.php
A range of lamps and lighting systems using WLEDs and aimed at promoting clean energy lighting to rural communities

Freeplay Foundation
56–58 Conduit Street, London W1S 2YZ, UK
Tel: +44 0207 8512616, Web: www.freeplayfoundation.org
Wind-up WLED Lantern using Freeplay technology. Freeplay also produces a foot-powered generator.

NEST – Noble Energy Solar Technologies Ltd
A-9, Aero View Towers, Shamlal Buildings, Begumpet, Hyderabad 500 016, India
Tel: +91 040 2776 2559, Web: www.solarnest.net
Manufactures and sells solar lanterns and other solar equipment

Sollatek UK Limited
Unit 4/5 Trident Industrial Estate, Blackthorne Road, Slough SL3 0AX, UK
Tel: +44 1753 688300, Web: www.glowstar.net
Supplier of the Glowstar lantern and other equipment

'Rural lighting' was last updated by Practical Action in October 2010.

www.ingramcontent.com/pod-product-compliance
Ingram Content Group UK Ltd.
Pitfield, Milton Keynes, MK11 3LW, UK
UKHW021822140426
5217IPUK00003B/40